Geopolitics of Memory and Transnational Citizenship

CULTURAL MEMORIES

VOL. 10

SERIES EDITOR

Katia Pizzi
Director, Centre for the Study of Cultural Memory
Institute of Modern Languages Research, University of London

PETER LANG

Oxford · Bern · Berlin · Bruxelles · New York · Wien

Geopolitics of Memory and Transnational Citizenship

Thinking Local Development in a Global South

Clara Rachel Eybalin Casséus

PETER LANG
Oxford · Bern · Berlin · Bruxelles · New York · Wien

Bibliographic information published by Die Deutsche Nationalbibliothek
Die Deutsche Nationalbibliothek lists this publication in the Deutsche Nationalbibliografie; detailed bibliographic data is available on the Internet at http://dnb.d-nb.de.

A catalogue record for this book is available from the British Library.

Library of Congress Control Number: 2017961055

Cover image: Zanglais Commune, 2014. Photograph taken by the author.

Cover design: Peter Lang Ltd.

ISSN 2235-2325
ISBN 978-1-78707-978-6 (print) • ISBN 978-1-78707-979-3 (ePDF)
ISBN 978-1-78707-980-9 (ePub) • ISBN 978-1-78707-981-6 (mobi)

Published by Peter Lang Ltd, International Academic Publishers,
52 St Giles, Oxford, OX1 3LU, United Kingdom
oxford@peterlang.com, www.peterlang.com

This publication has been peer reviewed.

Printed in Germany

Contents

Illustrations

Figures

Boxes

Tables

Acknowledgements

'Rale Kreyon W Make L'

As a Visiting Fellow (2016–2017) in the Centre for the Study of Cultural Memory at the Institute of Modern Languages Research (IMLR) at the School of Advanced Study, University of London, I benefitted from an incredibly supportive and inspiring environment. This is the place where parts of this book took serious shape to uncover the limits of the narratives of postcolonial studies. Along with the Institute of Commonwealth Studies, the Centre of Post-Colonial Studies (CPS), the Institute Latin Americas (ILAS), and the Centre for Integrated Caribbean Research (CICR), this stimulating environment of cutting-edge interdisciplinary research was perfect for a citizen of the world whose research interest spans from the Caribbean to the Gulf, passing through Europe. Thank you, Dr Catherine Davies, IMLR director.

Exploring the precise contours of decolonial thinking brings insight and clarity to the understanding of the continuum of geographies of memory within which immigrants travel. This powerful mechanism has rarely been explored in articulating comparative socio-political and economic solidarities among Caribbean communities abroad. I especially thank Dr Katia Pizzi, director of the Centre for the Study of Cultural Memory (CCM).

Along my path as an independent researcher, I have come across some incredibly generous people. I wish to pay tribute to the late Jean-Claude Fignolé (1941–2017), the 'poet of memory', former mayor of Abricots, where it all began.

The journey ends in Zanglais (see Figure 1.1), at Jardins sur Mer, home to the late geographer Georges Anglade (1944–2010), and the late Robert Anglade (1942–2016). May their visionary souls rest in peace. I am grateful to my husband Christophe Charles Eybalin for allowing me the time to write this book.

I would like to dedicate this book to the Caribbean, part of the hopeful Global South, in spite of quakes and uneven development, and to the beautiful people of Haiti and Jamaica, who have taught me more about the power of memory and civic engagement than any text could.

Abbreviations

AADH: Association d'aide au développement
ACFE: Association des Collinois de France et Extérieur
ACP: African Caribbean and Pacific Group
ACSHO: African Caribbean Self-Help Organization
ACVB: Association Au Cœur des Bidonvilles
ADENAC: Association d'Entraide Action Cœur
ADFH: Assoc. Développement et Formation en Haïti
ADHM: Association des Haïtiens de Martinique
AEH: Association des Etudiants Haïtiens
AFHAD: Association France Haïti Artibonite Desdunes
AFHSEC: Ass. franco-haïtien solidarité échanges culturels
AHDEL: Agence Haïtienne pour le développement Local
AIDE: Assoc. Insertion, Développement et Education
AIMF: Association internationale des maires Francophonie
AJDDA: Association de journalistes
AJN: Association of Jamaican Nationals
ALBA-TCP: Alliance Bolivarienne pour les peuples de notre Amérique-Traité de Commerce des peuples
AMAGA: Association des maires de la Grand'Anse
APDH: Association pour le développement d'Haïti
ARRR: Association Resettlement of Returning Residents
ASCG: Association Solidarité des communautés Guyane
ASEC: Assemblée de la Section Communale
ASK: Association San Komplex
ASOLID: Action de Solidarité pour le Développement
ATHAIF: Association des travailleurs haïtiens en France
ATHP: Association des Taxis Haïtiens de Paris
BMEO: Black and Minority Ethnic Organization
CACH: Confédération Assoc. Convergence Humanitaire
CADG'H: Comité d'Aide Développement Gressier-Haïti

CAOHA: Collectif des Associations Haïtiennes Alsace
CARICOM: Communauté et Marché Commun Caraïbes
CASEC: Conseil d'Administration Section Communale
CBEN: Confederation of Black Ethnic Minority Nurses
CBLC: Council of Black Led Churches
CHASO: Collectif Haïti Solidarité Oise
CIRH: Commission Intérimaire Reconstruction d'Haïti
CNUCED: Conférence ONU Commerce Développement
COGOP: Church Of God Of Prophecy
COPE: Caring Opportunity Partnership Equality
CORPUCA: Conférence Régionale Recteurs, Présidents, Directeurs d'Institutions – Caraïbe
COSEVBA: Comité Soutien électrification Vieux-Bourg Aquin
DCP: Document Cadre de Partenariat
DFID: Department for International Development
DGCID: Délégation Générale Coopération Internationale et le Développement
DSNCRP: Document de Stratégie Nationale Croissance et la Réduction de Pauvreté
DWIA: Derby West Indian Association
EFACAP: Ecole fondamentale d'application Centre d'appui pédagogique
ENEL: *An Nou Ede Lakay* (Let's Help Home)
EPA: Economic Partnership Agreement
EUNOMAD: European Network Migrations &Development
EWF: Ethiopian World Federation
FEDHE: Fédération de la diaspora haïtienne en Europe
FENAMH: Fédération Nationale des Maires d'Haïti
FFBJ: Facilitators for a Better Jamaica
FHAH: Association des Franco-Haïtiens et Amis d'Haïti
FHAHG: Fédération des Associations Franco-Haïtiennes
FIS: Association France Immigration et Solidarité
FORIM: Forum organisations solidarité internationale de migrants
FPMH: Friends of Princess Margaret Hospital
FSP: Fonds de solidarité prioritaire
GAFE: Groupe d'Action Francophone Environnement

GAMAH: Groupe d'appui médical à Haïti
GRAHN: Groupe de réflexion et d'action pour une Haïti Nouvelle
HAIDEV: Haïti Développement
HAR: Haitian Appeal Relief
HSG: Haitian Support Group
IDH: Indice de développement humain
IHSI: Institut Haïtien de Statistiques et d'Informatique
IRCOD: Institut régional Coopération Développement
JAMPRO: Jamaica Promotions Corporation
JDUK: Jamaican Diaspora United Kingdom
JOD: Jamaicans Overseas Department
KOREKA: *Kontak Rechèch Karayb* (Contact Research Caribbean)
LAKAY: Les Haïtiens de Bordeaux (Haitians of Bordeaux, France)
LKP: *Lyannaj Kont Pwofitasyon* (Alliance against Profiteering; Guadeloupe, France)
MBMB: *Memn Bitan Memn Bagay* (Same identity, same solidarity)
MFAFT: Ministry of Foreign Affairs & Foreign Trade
MGF: Marcus Garvey Foundation
MHAVE: Ministry of Haitians Living Abroad
MIIC: Ministry of Industry, Investment & Commerce
MINUSTAH: Mission of United Nations for Stabilisation in Haiti (2004–2017)
NAJ: The Nurses Association of Jamaica
NBP: Network for Black Professionals
NGO: Non-Governmental Organisation
OAS: Organisation of American States
OECD: The Organisation for Economic Co-operation and Development
OSIM: Organisation Solidarité Issue de l'Immigration
PAFHA: Plateforme d'associations Franco-Haïtiennes
PARDN: Plan d'action Relèvement Développement National
PNGRD: Plan national de gestion Risques et Désastres (National Plan Risks & Disasters)
PNP: People National Party
RDNP: Regroupement Démocratique National Progressiste d'Haïti

RGPH: Recensement Général Population et Habitat (Population & Housing Census)
RRFU: Return Residents Facilitation Unit
SCAC: Service de coopération et d'action culturelle
STATIN: Statistical Institute of Jamaica
TORA: Trelawny Overseas Relief Association
UAG: Université Antilles-Guyane (University Antilles-Guyane)
UBL: United in Building Legacy
UCDH: Union communautaire Développement d'Haïti
UHUK: United Haitians in the United Kingdom
UJD2F: Union des Jeunes d'Aquin et de France
UJEHG: Union Jeunes Etudiants Haïtiens Guyane
USLOAFRADES: Union des St Louisiens de France développement de St-Louis-du-Sud
UTSIG: Unité de Télédétection et de Systèmes d'Information Géographique
UWI: The University of the West Indies
VSO: Volunteer Service Overseas

INTRODUCTION

A geography of extra-territorial citizenship through memory

Political geographers' interest in the dynamics of space and place has long been the core defining analytical framework of the spatial patterns of immigrants through time and in relation to states. It is crucial to understand how cities offer a space for studying the dynamics of citizenship/non-citizenship, and how the challenges of increasingly diverse displacement are related to the capacity of migrants to adapt to the host environment. I seek to develop a contextual framework through which to revisit decolonial thinking on past dispossession and debt (colonial Saint-Domingue), present peripheralization (the Caribbean), and future uncertainty (post-quake Haiti). The key point of departure is research that explores migrants' associations that have sought to achieve unity through the use of memory. In fact, the geographies within which migrants travel are not virgin territories, and I examine the role of memory in the way Caribbean communities constitute new forms of belonging and citizenship across major European cities, illustrating the neo-colonialist processes of circulation of ideas and capital.

In my analysis of Haitian civil society (1986–2006) prior to the earthquake in January 2010 that thrust the country into the media spotlight, I argued that two elements were essential for understanding a singular historic trajectory: the significance and resonance of the Haitian Revolution, which is interconnected with both the American and the French revolutions in the memory of its citizens and beyond; and Haiti's vibrant popular culture of resilience. Haiti is a unique example of a nation born out of an indigenous revolution at the beginning of the nineteenth century. Nevertheless, it remains captive to an 'unequal relationship' with the great powers influencing its political and economic future, struggling with governments unable to achieve a national unity project. As a result, a conservative reading of the socio-historical Haitian context, based on demographic or economic indicators or events imposed from outside, such as the US occupation (1915–1934) and the 1994 US intervention, fails to do justice to a complex, changing reality. To examine this reality, it is necessary to examine successive migrations and their impact on the society of origin, which in turn necessitates a multi-layered analysis focused on the country of origin rather than on ongoing issues (such as integration/assimilation) in the host society. This latter approach neglects the state of origin and its policies concerning its citizens abroad.

This introduction examines contrasting memories that rest on power relations across the Atlantic. At an international UNESCO-sponsored conference in December 2004, 'Memory, Coming to Terms with the Slave Trade and Slavery', Jamaican Professor Nettleford's eloquent intervention on the challenges of knowledge, ignorance and silence in relation to the Haitian Revolution and the struggle against slavery stimulated my research interest in the development of transnational communities of Caribbean origin. His vision of a symbiotic relationship between academic and political activity points to the pertinence of a plural and mobile Jamaican culture as an essential element within a societal awareness of national development (Nettleford 1979). I take this focus on migrants' circularity – the sum of mobilities of a migrant population settled overseas – to analyse similarities and contrasts between neighbouring populations. This phenomenon is rarely studied in terms of associative practices, political engagement and the contribution of migrants to local development in their country of origin. This 'world in constant motion' (*Tout-monde*) between two societies marked by a process of *creolization* (Brathwaite 1971; Glissant 1993; Cohen 1998) is the starting point for looking at deterritorialized processes that force us to decentre the vision of a fragmented Caribbean in line with a Eurocentric continental logic of insularity.

Primary questions and hypotheses: Mapping mobility, locating transnational practices and articulating cooperation through a South-South framework

The interest in the complexity of migratory flows coming from the Caribbean, a region still affected by its colonial past, leads to an examination of the mobility of people, their dispersion on a global level, their settlement in European and American cities, and their contribution to the intensified flow of capital, all of which make the Caribbean one of the migratory spaces that is most closely interconnected with a global system, and an enormous transfer of human capital to OECD countries. The challenge is to discover how associative transnationalism takes shape. Which factor accounts for the differentiation between the Haitian and Jamaican associative spaces: the institutional framework of the host society or the complex politics of memory focused on communities abroad?

The *Ramphal Commission Report* (Thomas-Hope 2011) on migration and development indicates that it is the smallest or poorest states that register the highest rates of emigration of the highly qualified. This figure is higher than 50 per cent in Caribbean nations such as Guyana, Grenada, Jamaica, Haiti, and Trinidad and Tobago. This means that more Caribbean-born higher-education graduates live abroad than remain in their country of birth. This trend continues to intensify interconnected networks in a context in which the Caribbean continues to be perceived and conceived of as a fragmented space, shattered into a scattering of islands (formerly 'sugar *isles*'). This trend is demonstrated by the geopolitical sub-regions the area is habitually carved up into: Hispanic Caribbean (James and Perivolaris 2000), French overseas Caribbean (Burton and Reno 1995), Dutch Caribbean (Allen 2009), British Caribbean (Payne and Sutton 2001). It is clear that geographic maps are neither neutral nor 'true' representations of a place, but 'are often biased expressions of political construct or geopolitical and cultural asymmetrical relationships' (Massey 1995).

Mobility describes an evolved concept of transnationalism that acquires particular pertinence in my research. The academic debate and political institutional discourse have evolved from the perception of migrants as a problem for destination countries ('migratory pressure') and for the countries of departure ('brain drain'), to the idea of their becoming an element of a solution, as actors of development. Apart from the impact of transfers (remittances, immaterial exchanges), is it possible to conceive of a win-win strategy whereby host country and country of origin reflect on the politics of recognition of migrants? Hence, my initial questioning seeks to characterize this population and their different modes of solidarities and practices for a 'usable' geographic dispersion. I then question the way in which the 'geographic dispersion of a social group is utilised as spatial resource' (Ma Mung 1999: 89), and from looking at practices, I question two key elements essential to migrants' networks: social capital and trust (Faist 1997; Bashi 2007).

Beyond the existence of historical institutional networks among Caribbean islands and European cities, and the cultural and linguistic diversity of the Caribbean space, Doreen Massey's idea, reframed by the geographer Tracey Skelton, deals with the way the exercise of power structures the content of maps, and the difficulty of grasping the entire '*pan-Caribbean*' (2004: 1–9). A formalization of space such as a map, which is 'neither neutral nor innocent', means domination of the space [is an] instrument of power over this space and the people who live there' (Lacoste 1985: 11). In a cartography of exchanges and contact in the Caribbean Sea, a 'place of movement, of circulations, of back-and-forth' (Dalmace 2012: 12), where solidarity takes multiple shapes, we must reflect on the diversity of Caribbean experiences in migration. In this sense, in spite of 'a geographic fragmentation marked by the colonial history' (Daniel 2005: 134), the Caribbean we are referring to is a distinct geographic entity, the insular Caribbean, simultaneously a migratory space of increasing flux and a strategic zone (the Caribbean basin for some strategic Americans), marked by geopolitical rivalries that give the impression of simply dealing with 'islands mutually turning their back to each other' (Lechaume 2001: 270).

Although attention has been increasingly paid to migration as a means of development for the host societies, the policies of the country of origin deserve much better research and analysis. A constant evolution of measures by emigration countries towards their expatriates has been noted (Ma Mung 2009: 51), but de Wenden points out that there has been much less analysis of migrations from the point of view of the sending state, in contrast to the rich literature dealing with the host state and international relations (2010: 80), even when certain states, for example, apply a politics of 'attention and ties' towards their populations dispersed throughout the world (Dufoix 2010: 17).

Green states that 'The thinking on migrations gains by being comparative; nevertheless, most global issues remain largely confined in a national outlook' (2002: 123). My approach attempts to transcend the 'state-diaspora' opposition in order to propose a reflection on the relationship between civic engagement with politics and the relationship between state and migrants, putting the associative space of migrants of Caribbean origin at the heart of a participative engagement and a decentralized regional cooperation between Caribbean migrants and hosts. Thus, the possible link between migration, development and South-South cooperation allows us to question the contributions as well as the limits of public development aid (APD), the strategy of co-development (solidarity development) and the British approach via the DFID (Department for International Development) by obtaining a better grasp of the role of the associative structure within a decentralized framework. To what extent does an understanding of the local origin context lead to a better assessment of similarities and divergences in the strategies the Haitian and Jamaican states employ toward their communities living abroad? How do these become pertinent or insignificant in a long-distance migrant associative mobilization? Globalization gives the impetus for comparative approaches to open up parallels among different national societies, to ask questions and to go beyond the implications of the politics of integration that have too often been addressed in isolation (Lapeyronnie 1993; Byron and Condon 1996; Raymond and Modood 2007).

This raises two questions: how is the migrant treated as a true actor that is simultaneously 'spatialized and territorialized' (Di Méo and Buléon 2005: 31) and how can we bring together the institutional representative and the associative citizen actor? The work on active citizenship relative to multi-sited migrant engagement (Bertini et al. 2010) allows for a better framing of the following questions: how can a long-distance engagement founded on memory be explained as a tool of development, and how can transnational citizenship be useful in understanding an original brand of simultaneous participation in the host and origin country? By development, I mean the capacity for self-sustaining growth in terms of human, social, and financial capital, raised as a critical resource for national development. At the same time, how does migrant associative action, often reported as being influenced by the political and legislative national framework of the host society, inscribe itself in a dynamic that is partially or profoundly influenced by this framework? Finally, what contrasts can be discerned in migrants' lived experiences in North American metropolises (Foner 1978,1998; Goulbourne 2003; Grosfoguel 2007)?

In considering the evolution of strategies used by migrants of Caribbean origin residing in large European cities, new ways of 'being' and 'belonging' in the transnational associative space emerge. Albeit through different processes, a significant dimension of deterritorialization but also of re-territorialization – in other words, of transnationalism in migrants' practices in relation to politics at the international, as well as on a pertinent local, scale – can be observed. This makes possible, as a potential alternative to local development based on the migrant associative structure tied to the local authorities, the articulation of the triangular unit associative space-state-civic engagement.

Research objectives and methodology

What sets the Caribbean apart and justifies a unique, site-specific methodology? It is with this question in mind that I will attempt to apply a decolonial thinking conceptual framework (Mignolo 1993, Maldonado-Torres 2007, Antonio Benitez-Rojo 2013) to the field of geography, migration and memory studies. Benitez-Rojo explains that 'Caribbeanness is a system, full of noise and capacity, a non-linear system, an unpredictable system, in short a chaotic system beyond the reach of any specific kind of knowledge or interpretation of the world' (1996: 295). Yet, finding the right conceptual and methodological framework which accommodates multiple perspectives necessitates asking about the locus of enunciation from which the understanding subject perceives colonial situations. In other words, multi-locality is a way of re-mapping new geopolitics while applying pluriversality of knowledge to challenge conventional notions of history, geography, and economy. I am dealing with 'Globalectics', a horizontal way of undertaking not just literary work but a re-conceptualization of space, places and people. This perspective derives from an equidistant positioning whereby a point can serve as centre or periphery in a world of artificial boundaries. I adopt a comparative and pluri-disciplinary methodology that allows for an interesting analysis of different aspects of both Jamaican and Haitian communities as Caribbean people and in different European socio-institutional contexts. Through the intensification of an inter-polar and multipolar character between different places of settlement, ties are created among Haitian associations in Parisian *banlieues* or in Cayenne (French Guiana) that differ from those created among Jamaican associations in Walsall or Birmingham and the parishes of Trelawny or Portland in Jamaica.

The first challenge I faced was building an inventory of associations (Haitian/Jamaican). Through personal contacts in Paris and its *banlieues* (Essonne, Seine-St Denis), I located the bulk of the locally active associations and I classified them on the basis of their activities. I used original and secondary archival sources on Haitian migration to France and its overseas territories. This data was the basis of my master's thesis at the American University of Paris (2008). In the fall of 2009, as part of my doctoral project (University of Poitiers), I went on a field trip to Guadeloupe, Martinique and French Guiana. The purpose was to test transnational networks between families and associations in the Hexagon and its overseas territories. My focus on the associational activism of Haitians, whether in the French Antilles or the Hexagon, led me to question the motivation behind such dynamism, as well as to want to fill the gap in terms of theoretical shortcomings in the literature of migration studies.

Table 0.1 Interviews done in the Hexagon

	Hexagon	Strasbourg
Dates	2008/2010	June 2010
Respondents	35	3
Group meetings	6	1
Sex	H : 25 F : 5	H : 3 F : 0
Main hometowns	Léôgane, Artibonite Croix-des-Bouquets	Jacmel, Gressier, Léôgane

Table 0.2 Interviews done in the DFA (French American Departments)

	Guadeloupe	Martinique	Guiana
Dates	January 2010	February 2010/ June 2012	March 2010
Respondents	25	13	19
Group Meetings	3	2	3
Sex	H : 19 F : 6	H : 10 F : 3	H : 15 F : 4
Main hometowns	Léôgane, Aquin, Cavaillon	Léôgane, Artibonite Croix-des-Bouquets	La Colline, St Louis du Sud, Fond-des-Nègres

Table 0.3 Interviews done in the UK

	Birmingham	London	London
Population	Jamaican	Jamaican	Haitian
Dates	2010–2011	2010–2011	June 2010
Interviews	25	3	5
Group Meetings	3	2	1
Sex	H : 14 F : 10	H : 3 F : 0	H: 2 F: 3
Main hometowns	Portland, St Andrews, Trelawny, Clarendon	Ste Catherine, St Ann, Portland	Fonds-des-Nègres, Arcahaie, Aquin

Table 0.4 Interviews done in Haiti and in Jamaica

	Haiti	Jamaica
Population	Haitian	Jamaican
Dates	2010–2011/2014	June 2011
Interviews	25	15
Conferences	3	2
Sex	H : 14 F : 10	H : 8 F : 7
Main hometowns	Aquin, Abricots, Limbé Montrouis, Port-au-Prince Zanglais	Kingston, St Ann, Ste Catherine, Trelawny

During my interviews and group meetings (see Tables 0.1, 0.2, 0.3 and 0.4) in the United Kingdom, I first documented the different interpretations of key notions such as citizenship, ethnicity or association in order to establish clear and unambiguous terminology applied in both the French and the British contexts. I became aware of the frequent use in the British context of 'organization' rather than 'association', which is so common in the French context. For example, Jamaicans born in England are familiar with associations called Black Minority Ethnic Organizations (BMEO). In any case, from a legislative point of view, unlike in the French institutional setting, the right to associate did not exist at least up to the adoption, on 9 October 1981, of the law removing restrictions to foreigners' right to associate. I was concerned with understanding interviewees' experiences in order to better situate them in relation to the institutional context of the host society. It was important to treat these people not as minorities inserted into the society, but rather as 'mobile beings' travelling with their system of values, beliefs and memories. As part of the interview protocol (cf. 'Conducting Research under the GDPR: Legal Bases' 2017), confidentiality was always discussed at the outset in order to acquire informed consent and build trust with respondents. Interviews were anonymous and the interviewee names given are pseudonyms. Today some of these respondents are well settled in the country, and others are in the process of returning, or even in an ongoing circular movement. I had five group meetings in Birmingham and in London: with the Nurses Association of Jamaica (NAJ) (Birmingham, 8 December 2010), the Jamaican Diaspora of the United Kingdom (JDUK)

(London, 30 April 2011), and the United in Building Legacy (Birmingham, 6 November 2010). Another two meetings were held in Walsall with the Trelawny Overseas Relief Association (TORA), which accompanied us after the Jamaican Conference (St Ann Parish, 16–17 June 2011) to the parish of Trelawny at Deeside (18–19 June 2011).

Framing memory in a broader reflection on transnational citizenship and its geopolitical interlinkages between cooperation and development aid led me to adopt a pluri-local spatial methodological framework. The challenge of transcending the national and ethnic perspective so often adopted in migration studies arose when mapping mobility, citizenship, and diasporic and memory politics. The interlinkages between the increasingly important role of non-state actors (NGOs, migrant associations) in the (mis)management of war, conflict and natural catastrophes provided the continuum for the exploration of South-South cooperation. The fortieth anniversary of the Caribbean Community (CARICOM), of which Jamaica is a founding member, brought to the fore the issue of rapprochement within the Latin Caribbean space. Haiti has been a member since 2002 and is the only non-Anglophone and first Francophone member country. Long-distance engagement may help in the rethinking of migrants' associations as a legitimate and powerful unit of analysis deserving more academic attention.

Structuring memory, civic engagement and place

The book is organized into two main thematic parts. My focus is on the interaction between Haitian and Jamaican expatriate associative spaces and the municipal institutions of the homeland. The main research object is the Haitian and Jamaican associations and their roles in collective mobilizations, such as the events of June 2011 at Ocho Rios (Jamaica). I also examine to what extent statist structures contribute to framing the associative space and to giving shape to multiple initiatives that aim to develop the state of origin. Finally, I argue that the mobilizations reinforce a true sense of belonging based on memory. This leads to a sustainable model of decentralized cooperation at the regional level. The Jamaican case provides an opportunity to examine how this process unfolds from the perspective of the state of origin. From the perspective of a cross-section of the institutional framework, I examine the factors that can explain the relationship between the state and the diaspora (Chapter 6). Migratory and financial logic sheds light on differentiated aspects of the relationship between migrants and their origin states, a true collaboration within a statist Jamaican structure with entrepreneurs in the United Kingdom, or the expression of a simple recognition of the Haitian state in the case of Haitians based in French Guiana. Political activism is a form of mobilization that favours migrants' investment at home and abroad, and thus boosts their potential to craft a different form of citizenship 'from *here* and *there*' (Chapter 7). My final argument points to the pertinent intersections between mobility, post-colonial memory and state politics for new paths of development (Chapter 8). Consideration is then given to creative ways in which spatialities of memory will be of interest to academics and practitioners who seek a better exploration of memory to bridge social, cultural, political and tourism geography (Conclusion).

Part I. From the Caribbean migration space to the French-British context: Circularity, creolization and the spatial practices of belonging

Defining contemporary mobility puts in perspective socio-political and economic structures and dynamics that make sense in the 'permanent state of migrancy' (Deleuze and Guattari 1987: 52). Mobility forces us to rethink fundamental notions of citizenship and a sense of belonging that brings back memory in the form of practices that transgress or dissolve the boundaries of contemporary social codes. Such a process challenges the classical notion of state-citizenship and thus requires further research.

The question of the citizenship of immigrants puts in relation mobility and multiple loyalties on the one hand, and, on the other, the dynamics of solidarity and the improvement of the quality of life in the origin state. This refers to a fabrication of an 'active citizenship' (Bertini 2010). Such citizenship is defined in terms of participation and free circulation. Seen as a 'process in a dialectical relationship beyond its normative content', active citizenship allows us to question the classical meaning of citizenship within the French Republican model. I argue that a transnational framework based on the interaction between notions of memory and distance, mobility and citizenship, gives meaning to the growing intensification of associative practices, with some degree of independence from the institutional framework, whether French or British. From this perspective, the extent to which family and community links are developed over time forms a continuous relationship that binds space where such interactions exists. Such space becomes transformed into a transnational social space dependent upon a complex system of relationships characterized by 'a primary mechanism of integration based on reciprocity in small groups, exchange in circuits and solidarity in communities' (Faist 2000: 189). In addition, space carries a social meaning that goes beyond the notion of extraterritoriality to fit a spatially dispersed group of people. This is where the salience of homeland political initiatives and policy enacted by the host society come into play. These may partly explain the full emergence of transnational social networks. Given that this is the era of globalization and blurring state boundaries, the 'changing role of the state in conceptualizing different dimensions of citizenship', and varying understandings of citizenship condition migrants' transnational practices. My argument about citizenship shows a shift brought about by the recognition of migrants' strong ties to the homeland as opposed to the classical view of assimilation. Mooney (2008) correctly points to the low visibility of mediating institutions catering to immigrants in a French Republican framework that promotes a unified national identity based on citizenship and not on ethnicity. Yet, what does it mean to link memory to transnational citizenship to produce local development in the Global South?

Part II. The making of a transnational citizenship: A new dialectic between memory and civic engagement

I explore how spiritual engagement and entrepreneurial dynamism by migrants and transnational associations move from a process of top-down policy to an inclusive mechanism beneficial to their homeland's local development. The articulation of memory coupled with a sense of belonging to a territory are expressed through a detailed examination of several active associations. Understanding these complex place-based temporalities is critical to engaging with history and memory.

Narratives of remembrance and Caribbean lessons in post-earthquake Haiti

The geopolitics of memory needs much more detailed attention to explain the complex interrelation between a sense of belonging and long-distance civic engagement. What differentiates associational dynamics is the nature of the relationship of transnational communities to the origin state. The analytical perspective considers a broad range of contextual factors specific to the countries of origin beyond the Latin Caribbean.

The aim is to identify how different spaces are articulated by engagement and to demonstrate how memory is central to cultural geography studies. The various chapters demonstrate the need to make sense of practices of remembrance in places that carry different meanings. Geography and politics offer ways of drawing together memory and space to analyse how Caribbean communities constitute new forms of belonging and citizenship across major European cities. In the final chapters of the book, I consider current debates about the legitimacy of neo-colonial interference through NGOs. Based on secondary sources and field research conducted in Haiti, Cuba and Jamaica, I question how a Caribbean developmental shift through the prism of a Global South framework based on an active role for diasporic Caribbean communities is gaining momentum within a broader geopolitical and pan-Caribbean solidarity. Finally, putting in dialogue regions that barely interact in the literature is intended to motivate future studies on the emerging connections between memory, long-distance civic engagement and South-South cooperation.

PART I

From the Caribbean migration space to the French-British context: Circularity, creolization and the spatial practices of belonging

To annihilate a set of people, the initial step is to erase their memory. Their culture and their history are destroyed. Then someone else gives them another culture and recreates another history. Slowly the people begin to forget who they are and who they were. The world around them forgets them even more. In the short-term, it may be that history is written by the winners, but in the long-term ...

— Milan Hübl

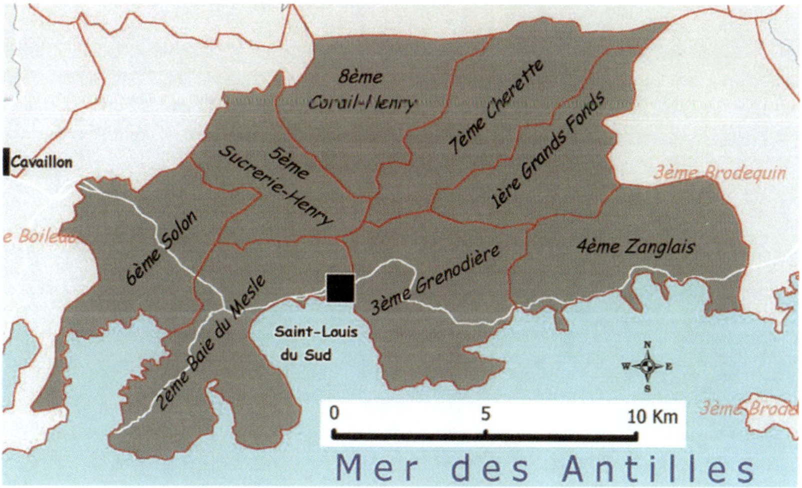

Figure 1.1 Zanglais in Cromwell Bay (1655).
Source: Joseph Vladimir Casséus.

CHAPTER 1

Encountering Caribbean memory in a pattern of dependency, independency and interdependency

It is tempting to think of memory as something that is fixed and defined within national boundaries. This chapter traces one of the main characteristics of Caribbean migration: the continuities in the mobility of the aboriginal people inside and outside the region. It demonstrates the historical meaning of this practice in the contemporary societal evolution of the whole region. In fact, this history of a region fully populated by pre-Columbian indigenous inhabitants remains under-researched, despite offering precious clues about a common regional pre-Columbian heritage. This matrix came out of a Caribbean population that was heterogeneous in religious creed, multi-ethnic, and linguistically and culturally diverse. Hence, I argue in the first paragraphs of this section that any informed analysis of Caribbean history necessitates starting with this era of the presence of the Tainos of Ayiti and Chémès of Quisqueya (the island of Haiti), which contextualizes an authentic *indigénisme* thinking that Haiti's intangible heritage holds. The island of Haiti (*Ayti Quisqueya Bohio*) carries various spelling of which, 'Ayti' (Fouchard 1972), Aiti (Dreyfus 1980), and more frequently 'Ayiti'. All refer to a territorial space. 'Hahiti' is composed of 'Ha', flower, 'Hi', big, and 'Ti', land, all of which signifies 'the island of mountains'. Fombrun (1992: 138) has suggested the adoption of 'Quisqueya' (in Creole 'Kiskeya') to replace the name of the entire island 'Hispaniola' decided in 1930 unilaterally by the USGB (United States Geographic Board) 'in homage to the Spanish colonizer, thereby erasing all traces of the Amerindians who occupied the island before the arrival of Christopher Columbus'.

My starting point is the exploration of the intangible heritage that is common to the cosmological matrix constituting the Caribbean as a living and diverse space. Prior to Columbus' arrival, mobility among aboriginal populations (Ciboneys, Arawaks, Tainos and Karibs) created a complex web of spiritual reservoirs. In 1492, the aboriginal population (called 'Indians') was estimated to be between 200,000 and 300,000 people as a result of exploitation in gold mines. By 1515 it had been reduced to 14,000, and by 1550 to only 500 on the island baptized by Columbus as 'Hispaniola'. This is the place where the first European military 'base' in the western hemisphere, La Navidad (*Nativité*), was created (Zinn 1980: 7). This place was previously known as the *caciquat of Marien*, a site of great archaeological research since Columbus' Santa Maria landed there. Columbus planted the very first cross in the Americas in the Bay of Conception in December 1492 (Saint-Méry 1796, re-issued 2004: 214).

Recent, independent studies (G. and D. Mathurin 2008; O. Wiener 1992; Laraque 1998; G. Valmé 2012; B. Bello, interview July 2011) have converged on the view of the primordial role of a wealth of immaterial Amerindian heritage in repairing the silenced past. In the autumn of 2012, a visit to the Museum of Guahaba (the Amerindian name of today's Limbé, in the north of Haiti) gave me access to an impressive collection of artefacts and elliptical stones that give evidence of the linkages between successive waves of Amerindians from the Venezuelan coast up to Yucatan, from the Lesser Antilles to Puerto-Rico, from Jamaica to Cuba, located at the twenty-second parallel North, at the bases of the triangle of the Bermuda triangle. The survival of the Amerindian practices throughout Caribbean culture as a whole is hinted at in the hypothesis of Berlitz (1975: 15), as well as the writings of Diego Valásquez de Cuéllar, governor of Cuba, and specifically his request that Hernan Cortes send to King Charles V the enormous bone fossils attributed to the possible presence of a 'gigantic population' with its own system of writing (Mathurin 1994). A peace treaty (Enriquillo Treaty) signed in 1533 between a Taino cacique Henri in Ayiti and a representative of the King of Spain is today recognized as the 'first international treaty in the Americas between Indigenous people and European' (Laraque 1998). While research continues, interesting initial findings reveal what may be fragments of writing found on stones that are inscribed with zoomorphic features. In Ayiti, *Roche à l'Inde* (Figure 1.2) presents a particularity, because it contains not only zoomorphic features, but also geometric features that closely resemble ancient scriptures of the 'Old World'. A comparison between the inscriptions of this petroglyph and earliest known forms of writing, such as the proto-Canaanite or Sumerian and Babylonian alphabet, leads us to question whether the history of alphabet has a missing link (Mathurin 2016).

Figure 1.2 Intangible Heritage: Roche à l'Inde, Limbé, Haiti. Photo by author: Limbé, Haiti (October 2012).

The singular nature of the discovery of geometric design in *Roche à l'Inde* points to a common matrix of spiritual practices among people of the Americas. Valmé's archeological fieldwork using the concept of 'axis mundi, a matrix of spatial planning' points to the island of Ayiti as a 'geographical focal point, a 'sacred land' designated according to a set of mystical beliefs' (2012: 7). He advocates more archaeological research based on a holistic territorial approach of the whole island. Going beyond the Treaty of Ryswick of 1697 whereby Spain recognized the right for France to establish itself in the western part of the island Haiti, the immaterial heritage makes senses out of 'a more holistic understanding of the whole community, transcending the geographical and political divisions and the alienation of caste and race' (Goveia, 1965: 337).

This runs counter to current analytical frames such as 'blackness' or '*noirisme*', the prominent propaganda ideology used against many in Haitian society of 1946, and especially by the François Duvalier regime, for divisive gains. Although my position is supported by authors such as Confiant (2012: 324) and Bulmer-Thomas (2012: 192), others (Nicholls 1979; Garrigus 2006) place too much emphasis on this superficial division and thus fail to capture the essence of a core ideal of liberty and freedom emerging from the aboriginal culture and being expressed through successive later revolts, up to the Haitian revolution. Against this backdrop, I argue that notions of circularity, memory and spatiality offer an analysis of the socio-political space and the construction of otherness tied to the transatlantic slave trade legacy.

The necessary return to the geo-historical sources of the island of Haiti through an interdisciplinary approach comprising archaeology, anthropology and, ethnology (Rouse 1939; Roumain 1941; Fouchard 1972: 128–130) shifts the focus onto unity through the preservation and promotion of the intangible heritage. Furthermore, mobility throughout the region is reinforced by a pre-Columbian pattern that demands the reconsideration of current climate threats (the sinking of small islands, the rise in sea levels) in a much longer perspective that includes the presence of the Taino and Karibs known as Garifunas, who are native to the Island of Saint Vincent and found in Belize and in Honduras (Sertima 1976: 10). Invited to speak at the American Congress before a commission on the celebration of the 'Columbus Quincentennial', Ivan Sertima's statements influenced the name of the commemoration to reflect not 'a discovery' but an 'extraordinary meeting and mixing of people of the world'.

The emergence of an extra-Caribbean migratory movement bringing Columbus was financed by the great European power of Spain. Profound changes would come to upset the space known as the Americas through a chain of diverse experiences, mixing identities and different ethnicities. The pertinence and intensity of the colonial plundering evolved according to the vitality of the economies of the 'metropolises', who were the beneficiaries of gold and spices. Some of the most significant examples are French cities such as La Rochelle, one of the most important ports for trade with the Antilles (Delafosse 1998: 7), Nantes, which supplied colonies with slave labour, and Bordeaux, which commercialized exotic products (Célimène

2002: 216). At the opening of the Musée du Nouveau Monde, the curator Delafosse gave a detailed account attesting to the significance of the trade between various French cities:

> The eighteen-century is the great era of the slave trade of La Rochelle given its 427 triangular voyages; less than Nantes, but a bit more than Bordeaux or Le Havre. There were peaks in 1738–1740 and 1783–1790. Then La Rochelle accounted up to twenty-two ships per year and a higher percentage in the total figure of illicit trafficking.

Delafosse stressed the contrast between a wealth of archival documents and the paucity of research publications'. In 1998, the exhibit 'La Rochelle, l'Afrique et l'Amérique' ('La Rochelle, Africa and the Americas') at the Musée du Nouveau-Monde coincided with the 150th anniversary of the abolition of slavery by the French state, on 27 April 1848. On this occasion, important research was done by the Archives Départementales of Charente-Maritime to retrace a route that was 'silenced, a slave trade [that] is a much too *rochelaise* business, led by a coherent group of big merchants, Protestants but also Catholics' (Delafosse 1998: 8). This story of the slave trade leads us to discuss the question of neo-colonial dependency, understood as 'the connection to one or more advanced capitalist countries, of a country nominally independent [Haiti] through ties such as those of a dominant country gets from deriving maximum benefits from a dominated country' (Joachim 1971), when an immigrant population comes from a formerly colonized country. In so doing, I reflect precisely on the nature of the linkage between colonization and immigration within a Republican institutional framework.

Mapping memory and mobility through circularity between ports

The field of political geography involves issues of how power interacts with space, how place influences memory, and how circularity creates and remoulds territory and economy. Socio-spatial and transnational theory helps to address new forms of geographies of memory politics, the politics of opposing memories, landscapes, and accumulation of power, and geographies of resistance. Yet, the place of memory, so central in revisiting the complex expansion of European influence through ports, remains under-researched. Idyllic destinations for present-day tourists conceal sites which once functioned as the bridgeheads of imperialist expansion, of colonization, and financial know-how of slave traders and merchants. My entry point is the circularity between the British port of Liverpool and the French port of La Rochelle before confronting the differentiated memory exhibited in La Rochelle's Musée du Nouveau Monde and in Montrouis's Musée Colonial Ogier-Fombrun (1492–1804). How these spaces allowed peoples and cultures to cross great distances to accumulate wealth exposes oppositional narratives. A recent report entitled 'Memory, slavery and reparation' (Frith 2016: 3) points to the persistent political will, at the highest level of the French state, to construct a 'shared memory to transcend a history that divided a majority of its population between masters and slaves, and to confront a criminal past after a long period of institutionalized silence'. However, this top-down process leaves little space for the construction of a common history which consists of fragmented and conflicting memories.

Box 1.1 Conflicting memory: La Rochelle and Haiti

'La Rochelle shares a common past with Haiti, through slavery. Let's not erase that, let's bet on the future by helping Haiti's development [...] The work of memory resides in the living together through respect and reconciliation'. This quote from a former French diplomat Daniel Groscolas reflects on the difficulty of articulating a programme of remembrance that has more to do with societal transformation and social justice for the youth than the usual official commemorations. He explains why it is pertinent for the local authorities to launch a serious work of memory in La Rochelle as part of the slave trade route. My first visit to La Rochelle dates back to September 2002. At the time, there was on display an important exhibition entitled '*L'esclavage dans les collections du Musée du Nouveau Monde, La Rochelle*', inaugurated by Dr Hippolyte, Honorary Consul of Haiti in Bordeaux, known as the 'Port de la Lune' (Port of the Moon).

> What caught my attention in particular were the numerous displays of family archives from La Rochelle dealing with Saint-Domingue (present-day Haiti); the profound economic ties between La Rochelle, as a slave port, and the Americas; and the lesser-known mystical dimension of the Columbus voyage to the triangle Cuba-Ayiti-Xaymaca (Taino chiefdoms). La Rochelle was presented as 'the port of the Templars', but also a major Protestant base. Yet, it seems like the remembrance of a slave port had been silenced for quite some time, before once more attracting some attention, at least with the effort of launching the exhibit.

Although I had visited Montrouis many times, I came back in June 2014 for a much longer visit of the Arcahaie region, where the Ogier-Fombrun habitation was situated. I had the privilege to speak with architect Gerard Fombrun, who in spite of his grave illness took the time to sign his famous book entitled *Pour que Renaissent le Prestige & l'Espoir* (2003) that recounts the ideals that inspired him. I remained very interested in how this *habitation* was the only sugarcane plantation in Haiti to be rehabilitated. I also observed several buses of students coming to the museum. This gave me the opportunity to test the educative dimension of this rare initiative not only to revive the nation memory through the museum, but also to do so in the context of a tourist attraction, Moulin-sur-mer.

Revisiting the slave route through conflicting mirrors

The idea of looking comparatively at British and French ports and their relation to the Atlantic world emerged out of a panel prepared for the Caribbean Studies Association (CSA) Conference of 2016 in Haiti. I put in dialogue two filmmakers, the Haitian Arnold Antonin and Jamaican Karen Mafundikwa. Antonin's *Haiti Apocalypse Now* questions the responsibilities for and failures that led to the catastrophe of 12 January 2010 with the magnitude 7.3 earthquake that hit the western department of the Republic of Haiti. The 20-minute film, documents the drama, courage and beauty of the Haitian people, as well as the many shortcomings of international solidarity. Mafundikwa's *The Price of Memory* explores the compelling narrative of the enduring legacies of slavery in independent Jamaica, and the case for reparations that reached both the Jamaican and UK parliaments. What is striking is how through this process the intertwinement of three institutions is revealed: The Bank of England's regulation on trade, the Church of England's collection of compensation money, and UK parliament acted on by the powerful West Indian lobby group (representing plantation owners) to retain slavery as a viable economic powerhouse at the core of Western industrialization for decades.

One illustration of this complex interplay is the rising importance of Liverpool as a centre for privateering during the Seven Years' War (1756–1763) and the American War of Independence (1775–1782). Records show that in 1778 no fewer than 120 privateers, carrying some 2,000 guns and crewed by close on 9,000 men, were sailing out of the port. In 1766, following appalling and ever-increasing loss of vessels (along with their passengers, cargoes, and crews) on the treacherous sands which surround the approach to the river, a Pilotage Commission was established to set up a Pilot Service. This would select and train specialists who would become Liverpool pilots. As trade with the Americas grew, so too did Liverpool's importance as Britain's West-Coast Gateway. In particular, copper was being imported from South America, while cotton was the main import from North America, to feed the demand of the growing industry of the Lancashire cotton mills. Wilberforce succeeded in abolishing the slave trade in 1807, but by this time Liverpool was firmly established as one of Britain's major ports. In *The Price of Memory*, Eric Lynch, an expert on slavery, makes an intervention in the history about Liverpool's place in that system, and questions where the wealth to build Liverpool's beautiful buildings came from.

> In Liverpool, most people ask when the beautiful buildings were built. The question is where did the money come from? Before Liverpool got involved in the slave trade, it was a small fishing village. Some buildings came from Mersey Docks & Harbour Board. The initials MD and HB are still visible. The site for their building was convenient to build docks needed for the slave ships. Liverpool became the richest city

> in the whole Europe ... On the other side, you can spot Martin Banks originally built on Hanover Street which eventually became Barclays bank. Two brothers Heywood invested heavily, built bigger ships and sold them to the United States, France, Portugal and Spain. Caribbean slavery generated taxes for British government.

This conversation raises the question of how the Liverpool slave trade fleet grew so rapidly. By the end of the century no less than 185 slave ships were operating out of Liverpool. The trade could be described as triangular, for the slave ships carried cargoes of tools and nails to trade with the West Coast slave traders for their captives. The second side of the triangle was the carriage of the slaves themselves (very few of the slaves were carried on to Europe), and the third side was completed on the homeward journey with the carriage of sugar, rum and cotton. No visit to the Port of Liverpool in the present day would be complete without a trip to the now obsolete South Docks or Brunswick System. Here today, in the Albert Dock, the visitor may view the superb architecture of eighteenth-century architect Jesse Hartley. The New York-based Restitution Study Group has uncovered evidence that two prominent slave trading bankers, Benjamin and Arthur Heywood from Liverpool, founded Heywoods Bank in 1773 on profits from the slave trade. Heywoods was taken over by Martins Bank in 1883 which itself was bought by Barclays in 1968. Historian David Richardson suggests that slaving and related trades may have occupied a third and possibly a half of Liverpool's shipping activity in the period 1750 to 1807. The wealth acquired by the town was substantial and the stimulus it gave to trading and industrial development throughout the north-west of England and the Midlands was of crucial importance. After abolition, Liverpool continued to develop the trading connections which had been established by the slave trade, both in Africa and the Americas. In sum, this intense circularity between British ports and the Americas testifies to the complex relationship between British industrialization and Caribbean slavery. This was especially clearly articulated in Eric Williams's *Capitalism and Slavery* (1944: 116): 'The purchase of the slaves on the coast of West Africa and their maintenance in the West Indies gave an enormous stimulus to metropolitan industry and agriculture ... Large profits were realized from the slave trade'. Additionally, Williams advocated for a decolonial process in education that had been 'sacrificed by the sugar plantocracy on the altar of labor-power availability and hence profitability' (St Pierre 2015: 66). In the case of Jamaica, successive waves of Spanish colonists in 1494, and then the British in 1655, dominated the slave trade for 200 years. Yet, the emancipation proclamation (1834) led to apprenticeship, 'a transitional expedient to convert slave labour into wage labour'. France, Spain, Holland and Britain set up the slave trades, but Britain shipped the most slaves from Africa and more.

The money went to England, the memory stayed here ...
The coffee went to France, the memory stayed here ...

A carrefour of triple memory

A port on the Atlantic, La Rochelle preserves traces of its past as an important maritime city. It has several historic sites such as the Town Hall, the gate of the Grosse Horloge, the famous towers of the harbour and the market. This description entices the ordinary tourist, who rarely knows about informative circuits of Protestant historical sites or of slave trade tied to La Rochelle. It is the second largest French slave port where many slave traders enriched themselves through trafficking slaves and the building of slave ships. What made the port of La Rochelle so prosperous? First, a mystical dimension envelops the place once called 'Port of the Templars', a Christian order of knights-monks, who depended directly on the Pope. Quickly the Order of the Temple became a wealthy and major power in Europe. The presence of the Templars at La Rochelle dates back to the birth of the city. From 1139, they received from Eleanor of Aquitaine watermills located at the mouth of the River Lafond as well as land in the centre of the city. 'This is where they established their commander, facing the port from which they were separated only by the wall of the Grand Bank' (Bonnin 2011: 13). Columbus' voyage to La Rochelle helped him to secure ancient maps that proved crucial to his adventure to the Americas. Second, the 500th anniversary of Luther's publication (1517) underscores the deep

Protestant culture through many touristic circuits ('maison du temple' in canton Bourgneuf founded by the Templars in 1224 under the protection of King Henry III of England). At the church Sainte-Catherine, lies the tomb of Eleanor of Aquitaine. Third, its less-known status as second busiest port involved in the slave trade (Deveau 2009) remains a challenging source for interconnected narratives across the Atlantic.

The Musée du Nouveau Monde, created in 1982 and inaugurated by the mayor of La Rochelle, Michel Crépeau, is located in a *hôtel particulier* of the eighteenth century, l'Hôtel Fleuriau. This hotel carries the name of the family who resided there from 1772 to 1974. It was built by the family Régnaud de Beaulieu in the 1740s. It was then purchased in 1772 by the ship-owner and rich planter of Saint-Domingue, Aimé-Benjamin de Fleuriau (1709–1787) before being rehabilitated some years later. The museum memorializes the slave trade. At the entrance of the building, a bronze plaque recounts the history of the building:

> Built between court and garden in a U-shaped plan connecting two lateral wings to the body of the main dwelling, this *hôtel particulier*, by its majestic and well-balanced calibre, by its exterior ornament and its interior decoration, inscribes in stone the social status of its owner whose wealth largely rests on the sugar production at Saint-Domingue (Haiti), the sweat and sufferings of African slaves, contributing to the prosperity of metropolitan ports. As with so many other buildings of the city, it reminds us that La Rochelle was one of the major French slave ports throughout the 18th century. Today, it is the residence of the Musée du Nouveau Monde, and a part of the collections refer to these sombre times of great maritime trade of La Rochelle and the tragic life of the slaves on the plantations.

The progress of the historiography sometimes leads us to revisit old works to shed light in new way. This has been the trend for some years now through the emergence of cultural and area studies in the Anglo-American countries and their set of operational concepts around race, gender, religion, and more recently agency (more difficult to translate in French as 'capacity for action'). This is the case in the attempt to uncover family ties between La Rochelle and Saint-Domingue (former Haiti). The journey of Fleuriau is emblematic of the complexity of the relationship between the two societies. Aimé-Benjamin Fleuriau left La Rochelle in 1730, aged twenty, after his father's bankruptcy in his native city caused him to join his American uncle in Saint Domingue. He got married, ignored his father's creditors, and lived lavishly as his fortune was quickly made through an enormous plantation in Cul-de-Sac. On his estate, there were a total count of 277 slaves. He died aged seventy-eight after attaining the status of nobility. His preoccupation with his slaves' health may have served another motive: to counter abolitionist sentiments in Paris with evidence of the benevolent paternalism of slavery (Cauna 1988: 221). *C*onsidering the relationship between families settled in La Rochelle and also in Saint-Domingue, of special interest is the way French families (particularly those from La Rochelle), confront colonial histories marked by the presence in France of Noirs and Libres de couleur. The salient characteristic of this treatment is the flexibility of their adaptation of conventional strategies in terms of parenthood, marriages, sponsorships and successions. The Rochelaise family is to some extent remodeled in light of the slave society of Saint-Domingue, but it always sought to preserve the interests of its members. It is necessary first to demonstrate, through a wide array of archives and iconographic sources, that the colonial traffic in general marks and shapes both metropolitan France, but also its colonies. Second, it underlines how family relations in an emotional and private mode inversely shape transatlantic exchanges and relations in the Creole society of Saint-Domingue. Glissant's unpredictable process of creolization with hybridization of cultures seems a more appropriate frame to grasp Fleuriau's case. Once his fortune was made, he was able to return to La Rochelle with children of a Creole mother, restore his reputation by paying off his father's debts, and obtain letters to support his accession to the nobility. The term *armateur* seems to fit, as it defines more appropriately the type of Rochelais merchants who engaged in slave trade: as both ship-owner and in charge of cargoes. In general, the profile varies from absolute owners to acting managers of ventures in which they held shares alongside other investors. Captains were responsible not only as commanders of ships, but also for carrying out trade on the African Coast. On 20 May 2015, a statue (Figure 1.3) in memory of Toussaint Louverture was unveiled in the courtyard of the Musée du Nouveau monde of La Rochelle. Ousmane Sow who designed the statue, also shows Toussaint reading the very first constitution of Saint-Domingue in 1801. This statue seemed to invite a turn to the other side of the Atlantic, to Haiti.

Figure 1.3 Red and blue Toussaint Louverture.
Photo by author.

'Et Haïti vint au Monde' (And Haiti came to the World)

Circulating memory becomes poignant when, contrasting with the epic history of 1804, one discovers how the Haitian people managed all by themselves to follow a tortuous road to liberation. Throughout the nineteenth century, Haiti was isolated from the international community. It was a century of great colonial empires being built and migratory waves all over the globe, with massive killings and massacres of aboriginal people. This was an era of slavery in Cuba and in Brazil, of the multiplication of 'coolies' in the British and French Caribbean, of the carving up of Africa by Europeans, Congo as King of Belgium Leopold's personal creation. This was also the century of the genocide of Amerindians in the United States of America, abuses against black people and the inhuman treatment of Amerindians in Latin America. The Haitian people did not arrive from Africa by swimming (Casimir 2004).

In 1804, as Haiti became an independent and free nation, the first and most powerful message was that it was the only victorious struggle against slavery that had also produced an independent nation. This achievement is symbolized at the entrance of the Musée Colonial Ogier-Fombrun by a wooden statue of the 'Nègre Marron', a reproduction of the bronze sculpture of Haitian architect Albert Mangonès. This symbol sets the tone of the efforts made to offer a place of memory. Three rooms are filled with collections of paintings, documents and objects tracing Haitian national history. A large wooden wheel used to crush the sugarcane reminds visitors of the main activity of the site. Very little is known about Frenchman Guillaume Ogier other than that, during the Haitian Revolution, he fled the plantation that had been running since 1760. Based on archives of Outre-mer de France, Ogier was the owner of the property of Montrouis from 1760 to 1799. According to a copy of his marriage act, he was married in Port-au-Prince on 16 August 1768. He died in Saint-Marc in 1799, during the Haitian revolution. His habitation covered 105 hectares planted with sugar cane and had a sugar installation. The indemnity paid to his heirs in 1827 was 50,300 *francs or* (Table 1.1).

Table 1.1 Living history at Museum Ogier-Fombrun

Main dates	Musée Ogier-Fombrun
1760	Exploitation of habitation at Montrouis by French colonizer (Bordeaux) Guillaume Ogier
1799	Death of Ogier in Saint-Marc

Main dates	Musée Ogier-Fombrun
1800–1804	As a result of the war of independence and the end of the colony and slavery, the habitation is abandoned
1804–1977	Exploitation of the land by several farmers of the region who became owners
1977	Acquisition of the ruins by the architect Gérard Fombrun
1977–1993	Start of rehabilitation programme of the habitation
1993	Official inauguration of the Museum Ogier-Fombrun

Source: Author, based on Fombrun Collection, 2003.

In one of the rooms where the historical patrimony is displayed, the Musée Colonial Ogier-Fombrun recounts the depth of the sugar industry and its necessitating a highly labour-intensive mode of production. Oruno D. Lara, Guadeloupean-born historian has contributed to show the influence of the Haitian Revolution in the Caribbean, the Americas and beyond: 'While the trans-Saharan trade practised by the Arabs from the seventh to the nineteenth centuries in Africa itself, in the Mediterranean and in the Indian Ocean involved some twelve million people, according to recent estimates, the transatlantic slave trade is thought to have deported some fifteen to eighteen million captives'. For example, in 1789, French imports from the colonies amounted to 185 million pounds, against the 78 million pounds from exports. Thanks to the re-exportation of colonial products, France maintained a favourable commercial balance to a trade based on the mercantilist expansion, as 'colonies were created by and for the Metropolis' (Lara 1992). The French presence since 1635 under the reign of Louis XIII was initially made of pirates and buccaneers 'frères de la côte' settled on the island of La Tortue, 'separated from the neighbouring land (today Haïti) by a natural canal and regarded as more fertile land than the island of Ayiti' (Funck-Brentano 1979: 8). Two types of maritime circuit were developed from the French Antilles to the metropolis: one concerned the triangular trade of the West African coast, around the Gulf of Guinea, and the other directly linked the Antilles to the ports of the Atlantic façade. Slaves were used as capital for production while simultaneously considered merchandise on the slave market (Salas-Molins 1987: 179).

In this context of specific economic exploitation, two major obstacles favoured the reconsideration of the slavery system: the limits of industrialized production due to land concentration and the scarcity of technical means, as well as social tensions. As Lara (2004: 55) puts it, 'The rebellion of the slaves on Santo Domingo, in August 1791, sparked off a general insurrection which led to the abolition of slavery and to the War of Independence. It marked the beginning of a triple process of destruction of the proslavery system, the slave trade and the colonial system'. Haiti's independence in January 1804, which occurred in the middle of the Napoleonic Wars, was seen as a threat by all those countries still engaged in the slave trade (Bulmer-Thomas 2012: 161). It continued to challenge the collective intellect in engaging a memory of the trans-Atlantic slave trade and plantation slavery. It forced French traders to redirect their commercial strategies and to look for new plantations in other parts of the world (the Mediterranean Sea, the western coast of Africa and Asia). The specific demographic characteristics of the French Antilles greatly influenced conditions for production and the evolution of social structures (Cottias 1998: 3).

The decline of the great sugar production, multiple peasant rebellions such as the ones of 1843 in Haiti and of 1865 in Jamaica (Sheller 2000: 35), the development of small peasant sharecroppers and the isolation of small villages constitute the major characteristics of this period. Former slaves, eager to establish their own autonomous communities away from plantations, settled on free lands mainly in Haiti and in Jamaica. Others who could not survive without their occupation of seasonal farming, moved to work on plantations, especially in Cuba and the Dominican Republic. The Haitian presence in Cuba, estimated at 600,000, dates back to the beginning of the twentieth century, between 1902 and 1913, then in 1913 and also in the 1930s. They settled over several generations primarily in the eastern part of the island, in the city of Barranca between Santiago de Cuba and Holguin (Hume 2011: 76). Originally from Port-de-paix, Cayes and Plaine du Cul de Sac in Haiti, they arrived to Cuba for seasonal agricultural labour, and then remained there. The unique feature of this migratory pattern is how these communities managed over the years to preserve a distinct identity fuelled by the memory of the Haitian revolution, as well as spiritual beliefs and cultural practices through societies called '*Sociedades de Tumba Francesa*', recognized by UNESCO in 2003.

However, in the case of Jamaica, after abolition in 1834, an economy of plantations in crises caused a great rural exodus, and some settled in Port-au-Prince as mechanics and coachmen (Depestre M. 2004: 58). Starting in 1850, after the abolition of the slave trade in the French colonies, Guadeloupeans and Martinicans came by the hundreds to Haiti, opening shops. A law of 6 September 1860, under the government of Nicolas Geffrard, freely granted 5 hectares of land to a whole family of blacks coming to settle as farmers in the republic (Joachim 1979: 101; Smith 2004: 35). These historical facts shed light on a rarely disclosed immigration that could explain the complexity of 'negritude', a matrix of a space always in movement. 'It became difficult to count the great deal of Haitians whose ancestors, originally from the Antilles under French domination, migrated to their new homeland at the end of the 19th or beginning of the 20th century' (Price-Mars 2009: 54).

Along with the arrival of labour from Asia, Africa and Europe to respond to the changing economic situation of the post-slavery era, multiple intra-Caribbean flows emerged. During the second phase, from 1885 to 1920, the Caribbean experienced a huge exodus: the first North-American sugar companies invested massively in Cuba and Puerto Rico, then in the Dominican Republic, by partially mechanizing the treatment of sugar canes. 'The workforce no longer recruited for more than five to six months of intensive cane cutting began to constitute the first seasonal migratory waves of workers forced to return to their islands of origin once the harvest over' (Domenach 2002: 16). These intra-regional movements emerged out of specific development projects conceived by capitalistic circles, such as the building of the Panama Canal, which attracted a great deal of people from the Caribbean. 'From 1904, there are approximately 20,000 Barbadians, 5,500 Martinicans and 5,000 others coming from all the small Antilles, as well as more than 20,000 from the British Antilles, who are documented as having died in this project' (Domenach 2002:16). This phase is characterized by great economic growth tied to a specific petroleum sector in Venezuela, in need of a qualified workforce from the British and Dutch Caribbean islands.

Hence, displacements followed the investment of foreign capital as European and American enterprises developed their activities in the region (Thomas-Hope 2002: 15).

The Caribbean basin gradually became the site of a new sociological space in which social relations and very particular family ties were at the core of migratory dynamics (Casimir 1991: 99). Socio-economic mutations accelerated with the impact of a determining factor in the geo-political equation in the Latin Caribbean: the emergence of the United States of America as an imperialist nation intent on controlling this other Mediterranean body that enclosed the Panama Canal. In line with the Monroe Doctrine, its economic expansion was expressed through multiple cases of interference in Central America and the Caribbean (Audebert 2009: 19). This influence took the form of a series of military interventions and occupations, along with North American financial capital causing a great intensification of labour and mobility, as well as multiple resistances and peasant revolts (Gaillard R. 1973, 1982, 1983; Castor 1987/8).

Cuba was under US occupation, by the American Marine Corps, between 1906 and 1908, as was the Dominican Republic between 1916 and 1926. These two events greatly contributed to the flow of clandestine Haitians to both countries. For example, in 1902, Haitian migrants followed the different phases of the rise and fall of North American capital to Cuba, the Dominican Republic and finally to the United States and Canada. As a result, the US occupation between 1915 and 1934 consolidated a growing role for Haiti as a main source of cheap labour for the American sugar industry, which was in full expansion (Gaillard 1990: 137).

Between 1920 and 1940, relationships of political dependence with the United States of America deepened. The Caribbean diaspora is an expression of a long and ongoing colonial and neo-colonial exercise, as well as of responses to factors of internal instability (Trouillot 1983: 216). This was an era of internal instability exacerbated by American interferences in the Caribbean and Central America region. This period marked by the US occupation in Haiti (1915–1934), resulted in the first major migratory intra-/extra-Caribbean flows.

The Caribbean became a region whose strategic interest influenced American foreign policy even when certain countries were culturally and politically more orientated towards South America and other regions (Lewis 1984: 6). The framework of the Cold War imposed a geopolitical dynamic in the Caribbean that was very much conditioned by relations between the United States of America and Cuba. While the implications of the new development of 17 December 2014 will be discussed later in this book, a general trend may be noted. Indeed, in a movement of strong dependency, some states of the region sought diverse degrees of autonomy from this powerful neighbour (the Jamaican government of Manley in the 1970s), while others preferred to cultivate close relations with the USA. These two orientations express the policies of Caribbean countries in line with the perception of the interests and needs of their populations. In the Jamaican case, there was a social movement before the Second World War, as the Jamaican Marcus Garvey mobilized thousands of people in the 1920s around an 'active racial and social awakening' (Guerin 1956: 121). The economy of this island, just like others, would continue to be affected by this situation of dependence:

> The banana industry, which was first introduced in an effort of diversification, later became the main product in the economy of some islands. The removal of the preferential tariff caused national product (GNP) to fall and led to severe costs paid by the rural population. This forces us to conclude that the future of the Caribbean cannot depend on the exportation of raw materials. (Interview with former Jamaican PM P. J. Patterson, Kingston, 14 June 2011)

The incorporation of migrants into each receiving country shows new ways of life, and shapes identities and the relationships of the self with the other. Many Caribbean people spread around the world, and in particular to former colonial cities and such displacement was also the result of a demographic explosion of the urban population of some islands (the rebellions of 1938 in Jamaica, the labour union movement in Trinidad and in Haiti from 1946 to 1948). In respect of the societal and cultural mutations of the contemporary world, this phenomenon of 're-diasporisation gives a particular pertinence of paradigm in the Caribbean, putting it at the core of a revolution in the way to rethink identities, ethnic and cultural solidarities, as well as collective memories' (Cohen 2002: 6).

In 1982, the Caribbean intellectual Sony Ramphal prided himself on a trans-Caribbean culture and perceived a growing phenomenon of transnational regionalization. He called for the 'full and free development of the people of the region with the creation of a sovereign nation state of the Caribbean uniting the entire archipelago' (Manigat 2007: 709). Faced with serious challenges in the twenty-first century, the Caribbean idealized as 'one' remains deeply marked by a growing presence of its population abroad (high- and low-skilled workers), and it is searching for an interdependence that can hardly exist between unequal powers while metropolises and peripheries continue to engage in a relationship of dependency (Williams 1975: 532). Rethinking memory is pertinent on two aspects: the interdependence of Caribbean territories of the Commonwealth and of Francophonie and its overlapping of a diverse Creole world, and the role of transnational circularity in terms of ruptures and continuities in the practices of belonging with the rest of the Global South. In this sense, Knight and Palmer state:

> European intrusion interrupted abruptly the original structure of an historical development of indigenous societies of the Caribbean islands. It severely altered its physical environment and diversified its nutrition, complicating its epidemiologic systems, producing new chains of microbes and inextricably tying them to a broader world beyond the Atlantic Ocean. Since, they remain attached to it. (1989: 1)

Conclusion

This historical chapter on migratory waves in the Caribbean, profoundly marked by mobility even before the pre-Columbian era to become a space of forced labour (the slave trade, indentured labour), demonstrates the pertinence of the continued influence of a common colonial memory with economic and geopolitical challenges in the Caribbean. Under such an influence, the extent to which circulations between colonies and metropolises are embedded leads to a more detailed and specific discussion about the orientation towards two major destinations: France and the United Kingdom.

As I move to the space and place, my attempt is to put in dialogue transcontinental connections of one history, revealing both a narrative of predatory economics and intentional silence, but also a cartography of resistance and resilience symbolizing an ideal of true freedom for all. Memory is colliding without necessarily being in true dialogue. In Chapter 2, I will inscribe these migratory dynamics into the European context and, in particular, these two countries, marking the third phases of this historical link between migration and (de)colonization in the second half of the twentieth century.

CHAPTER 2

Decolonial mapping of Haiti and Jamaica: Memory, labour and mobility

This chapter situates its discussion of decolonial memory in the context of Haiti and Jamaica. It draws on the evolution of migrants' strategies in the host countries and their socio-economic and political impact on the states of origin. Despite different migratory histories, the populations of Caribbean origin who settled in the United Kingdom and in France share a large number of similarities in terms of residence as well as professional integration. Such commonalities are the expression of a common experience within Caribbean emigration. By addressing collective remittances, capital and mobility, I wish to highlight the ways in which memory is embodied in the proactive approaches adopted by the Haitian and Jamaican governments in working in partnership with its citizens abroad. In this respect, the potential of transnational citizenship fueled by a sense of collective memory as a policy lever is explored in terms of a coherent policy for local development. I draw on statistical evidence to show the evolution in the reorientation of fluxes linked both to socio-political events (in the countries of origin) and at the legislative level.

Decolonizing memory starts with a redefinition of what memory is and why it matters in the narratives of Caribbean historiography. I begin this chapter by briefly describing a context of marronage common to both Haiti and Jamaica, and beyond. Marronage, or the act of becoming a fugitive, has long been seen among historians as an innate desire for liberty, especially in the case of Saint Domingue, a former colony of France. The first migratory phase up to 1885 was a new era marked by a period of strong mobility towards the mountains and frequent occurrences of marronage towards inaccessible regions. In Jamaica, migration became a form of resistance, a way to escape from desperate poverty

(Beckford 1982: 53). Jean Fouchard (1981) argued that the growth of marronage provided the foundation for the slave revolt in 1791 by providing both a large number of potential revolutionaries and by helping to forge leaders capable of organizing an armed resistance effort. Sabine Manigat (2009: 314) has established a pioneering typology of the female role in the overall construction of the Haitian nation (1790–1804). By revisiting embodied memory on the part of some exceptional female warriors, I retain the category of '*femme actrice*' (female actress) symbolized by autonomy, leadership or being an inspirational figure, and the impact of the action.

> In the Caribbean, a Jamaican has little chance of meeting a Guyanese or a Barbadian outside privileged circles of literature, of cricket or of the regional university. It is therefore in 'exile' that people from the Caribbean islands discover common features and forged ethnic ties which do not exist in the same form as in the origin states. (Goulbourne 2002: 71)

> We, Jamaicans and Haitians, are neighbours, but nonetheless ... there is a colonial ignorance, a fragmented knowledge and no serious regional agenda to reflect together about global issues such as those deported. (Interview with Dr Carolyn Cooper 22 June 2011, Kingston)

Haiti and Jamaica share a very long history that has its roots in a common geological formation of these islands, which detached in the Myocenic era. At the conference 'The physical geography of Haiti' held by the Society of History and Geography of Haiti (18 June 1933), academic works traced the complexity of great structural features of the presqu'île du Sud of Haiti that are similar to the geology of Jamaica. Nevertheless, little research has dealt with migration since the Spanish era (Haiti: 1492 to 1630; Jamaica: 1494 to 1655) attesting to the existence of an indigenous population composed of islanders speaking the same tongue as the marcorix. These two lands witnessed French-British rivalries since the settlement of privateers at l'Île de la Tortue and at Port-Royal around 1630 (Ehrengardt 2009: 181). Certain legends have fuelled the common Caribbean imagination on slave maroon and their battles in the mountains against the British powers (the presence of Cromwell in 1655 in once called Cromwell Bay, today known as Zanglais, see Figure 1.1), as well as their 'proto-peasant adaptations'

(Trouillot 1998; Besson 2000, 2016). This led in 1739 to the first peace treaty of international relations between a colonial power and its colonized population (Henke 1993: 21). Some research on the political and judicial system of the 'Jamaican maroon communities' reveals a complex autonomous system vis-à-vis the rest of the Jamaican territory at the time (Price 1983a: 15; Bilby 1992: 55). The slave maroon and the rebel Dutty Boukman from the parish of Saint Thomas arrived at Saint Domingue and led the first victorious slave insurgency of 1791.

> Jamaicans, once informed of the saga of this epic event in the history of a people of the Americas, also learned that Haitian authorities refused to return fugitive slaves by declaring that they will send back boats with slaves knowing that these slaves from there on were considered 'citizens' of Haiti and thus free with a sense of right and duty, part of a society of proud human beings. (Nettleford UNESCO Conference 2004)

There are no records in the archives about Dutty Boukman's exact age. He may have lived to around thirty-three (circa 1760–1793). He was born in Jamaica and sold by his British owner to a French planter from Saint-Domingue. Celine Fatiman worked with him as a warrior born to an African slave woman and a white Corsican man and was sold as a slave to the French colony of Saint-Domingue. Celine may have been a descendant of Manding origin (known to be rooted in the Kingdom of Mali in West Africa). Recent research (Clérismé 2009; Bernard 2010–2013) traces the presence of Hebraic heritage in the Southern part of Haiti (in the commune of Coteau: Habitation Jerusalem, Morne Sinaï, Sinaï, Zizalem) and of Islamic heritage in five localities in the Northern part of Haiti (Camp-Louise, Balan, Maïssade, Port-Margot and Pignont). One of the practices that reflect such immaterial heritage is a funeral ritual which endures to this day. Fatiman took a very symbolic role in the Haitian revolution as a spiritual figure who participated in the key ceremony at 'Bois Caiman' (myth or fact contested by Leon Francois Hoffman, 1993; Beauvoir 2002). She is not known to have fought with machetes in any battle; however, she remains respected for her mystical power to help break the chains of repression. She became a first lady through her marriage to the seventh president of Haiti, Jean-Louis Pierrot, who ruled between April 1845 and March 1846.

Feminist scholars have long explored multiple and contested geographies of memory through analysis of historical events. From Zabeth in Haiti to Nanny in Jamaica, raising the issue of whether and how collective memory should be instituted is a complex process in popular patriotic culture. For instance, in the writing of Jamaica's history, many historians have recorded the epic deeds of Nanny during the eighteenth-century Maroon rebellion. In her own recovery of the role of women in Jamaican society, the historian and stateswomen Lucille Mathurin-Mair dubbed Nanny 'the Rebel woman', a term with particular resonance in the contemporary popular cultural setting of reggae, dance hall and rebel music. Nanny was attributed special powers in the maroon resistance to British enslavement: she was deemed the major strategic mind and imbued with the capacity to repel bullets by 'bouncing them off her backside'. It is unclear whether the latter point was real or imagined. It is possible that, despite the nominal leadership of Cudjoe, it was Nanny herself who was the superb tactician who, through the force of her personality and courage, won the loyalty and following of the group. In her own lifetime, Nanny received 500 acres of land from the government of Jamaica for herself and her people. In the 1960s, with the emergence of the nationalist movement and the need for national heroes, Nanny was commemorated as the first National Jamaican heroine. It seems that, with time, she has also achieved another status in Jamaican society.

The evocative power of female warriors such as Zabeth is a way to stage the past within Haiti but also abroad in the dispersed communities. She is the name of a rebellious slave who represents the irreducible maroon who is whipped and tortured after the first attempt at escape. She symbolizes the slave whose ear is cut off after her second attempt. She is again the one whose arm they cut off when she is caught after her third flight. She remains the resilient figure who persists in her defiance, accepting the punishment of a loss of a leg in her running away again for the fourth time. This is the extreme individual case of instinctive refusal of colonial slavery, a case of indigenous insubordination that Dr Price-Mars will qualify as 'a long-lived love of liberty' (Manigat 2004). This is a mysterious tale that keeps coming back in the journals of 'habitations' held by colonizers of Montrouis, at Fort-Dauphin (today Fort-Liberté), in the area of Lake Azuei, where a spring named *source* Zabeth carries the same legend.

Not all stories are fictional, but some are imbued with elements of fiction which may have been necessary for the purposes of the time. With this in mind, I will reflect on the acts of remembering of three figures often bypassed while the roles of their companions are magnified. The first is Victoria Montou, known as Toya, a slave of the habitation of Cahos and a loyal companion in arms of Jean-Jacques Dessalines. He is the key army general who organized the decisive 'Congrès des Clairiere' (14–18 May 1803) in Arcahaie where the army of rebels known as 'armée des Incas' united under a common flag (Supplice 2017: 4). Very little is known about Toya's upbringing other than her fierce courage in carrying out humiliating and hard work on the plantation. At the Déluger habitation where she was later transferred by her master, Toya was in charge of approximately fifty slaves. Her voice was strong and she was respected as a commander-in-chief. During the agitation and period of rebellion, her group of insurgents was recognized and oppressed. Toya fled as two soldiers came after her. She resisted and was finally held captive. In 1805, Emperor Jacques Ier (Dessalines) called his doctor Jean Baptiste Mirambeau to the rescue as Toya's health became critical. He said: 'This woman is my aunt, please take good care of her as you would have done to me. She had to face hardship just like me, all the pains, all the emotions during the time we were both condemned side by side on the plantation'. Toya died on 12 June 1805 and was buried the next day. This also shows the gratitude the Emperor had for Toya, the relentless and brave warrior who naturally acted as a true freedom fighter.

The second is Marie-Claire Heureuse Guillaume Félicité Bonheur Dessalines (1758–1858) who remains a 'national treasure' and the embodiment of piety. Born in Léogâne, Marie-Claire Heureuse has often been compared to Queen Anacaona who was also born in the former Yaguana (today Léogâne). Anacaona in Marcorix means 'Flower' ('Fleur dorée'). She is recorded as the first poet of the island of Hayti Quisqueya Bohio (Haiti). She was the sister of Behechio, the queen of Xaragua and the wife of Caonabo, cacique of Managua, the most powerful chiefdom (caciquat). Within Managua, there was the rich province of Cibao and the River l'Artibonite. Marie-Claire Heureuse's sense of purpose was dedicated to the glory of God. Her Christian life indicates her vocation as the wife of the founder of the nation. This account is based on the historian Marceau

Louis's book published in 1954, to coincide with the 150th anniversary of the nation's independence. Marie-Claire Heureuse was born in 1758 and raised by one of her maternal aunts, Elsie Lobbelot, who was the governess of the monastery of the Saint Dominican order at Léogâne. In 1795, she was widowed upon the death of her first husband Pierre Lunic, who was in charge of the habitation of Frères de Saint Jean de Dieu. She then married General Jean-Jacques Dessalines on 2 April 1800 in the city of Saint-Marc. On 6 October 1804, Jean-Jacques Dessalines became emperor under the name of Jacques, Emperor Ier by the religious figure of the Capucin Order, Corneille Breille. Mme Dessalines became empress. An imperial court was created. She had never given birth; however, she adopted Dessalines's children, Célimène and Innocent, born in Saint-Marc. On 1 January 1805, the first anniversary of national independence was commemorated at the Imperial Palace. Victor Schoelcher recalls: 'It is at Saint-Marc that Madame Dessalines retired and I was given the honour to be introduced to her though the generosity of General Bonnet and his son. Madame Dessalines is a tall "négresse", with a calm and pure face, and of great beauty. She was wearing white as she usually did. She was living peacefully and with dignity away from the unstable period after her husband's assassination on 17 October 1806'. During the night of 8 to 9 August 1858, she died at the age of 100 in the city of independence, Gonaïves.

The third portrait is Suzanne Louverture, known as a talented farmer who continued to grow mainly coffee even after her marriage to François Dominique Toussaint Louverture. She was reluctant to actively participate in his public engagements. She is remembered for her generosity in sharing her wealth with the poorest. During all the years of torture, she kept the same line: 'I will not talk about the matters of my husband with his persecutors'. On arrival in prison, she weighed 250 pounds. When she left France, she weighed 90 pounds. It was a Suzanne mutilated, devoid of all her nails, with several broken bones, who went to Jamaica where she would die on 19 May 1846, at the age of sixty-seven. I have presented some key female warriors in Haitian historiography that contributed to the success of Haiti's independence. These tales reinforced part of the consciousness of a culture of resistance passed through oral history to overthrow relations of exploitation.

From this revolutionary context, there are documented exchanges between Haitian authorities and the British governor, who, in 1728, brought Arabica coffee seeds from the island of Haiti to the *Blue Mountain* of Eastern Jamaica. An exodus of Haitian coffee growers would follow, escaping the Haitian Revolution towards Jamaica. Some Jamaicans migrated to Haiti in irregular waves from the end of 1880. The British consul in Port-au-Prince recorded that approximately 2,000 Jamaicans resided in Haiti, some working for British companies in the north of the island (Ile de la Tortue), attracted by the Haitian economic boom of this period (Smith 2004: 34). Such movement back and forth between both islands also involved former Haitian presidents who resided in Jamaica, in particular between 1844 and 1944.

While the slave trade was banned in Denmark (1803), Great Britain (1807) and the Netherlands (1814), the steady increase in the demand for human labour did not abate, due to the magnitude of sugar and cotton production, which in turn generated a rationalized migratory movement on a great scale. In less than a century, hundreds of thousands of indentured people, often forced to emigrate, came from Asia (mainly India and China) (Klein and Singaravélou 2012: 24). The Congress of Vienna in February 1815 recommended banning the slave trade on behalf of each participating country, including France and Portugal, but a meaningful total abolition was rarely observed (Unesco 2004: 15). The abolition of slavery in the United States of America only came about in 1865 (with the thirteenth amendment of the constitution). Even when the British government banned the slave trade in the Atlantic Ocean at the beginning of the nineteenth century, the 'slave trade remained alive in Cuba until 1868' (Domenach 2002: 14).

The great fluidity of the movements of people from Jamaica and Haiti is characterized by two marked features: cheap agricultural labour and a range of destinations. Since 1868, when Cuba was considered to be the first global sugar producer, labour coming from Haiti and Jamaica in particular was enormous (Marshall 1982; Levison 1987: 97). Similarly, a great number of middle-class Jamaicans would move to Central America, where they worked on the construction of the Panama Canal, as well as in Costa Rica, Nicaragua, Honduras and other regions. Candis Hamilton describes in detail such Jamaican migration, of which there is little archival documentation (2006: 405–411). Some patois can still be heard in certain parts of these countries. Other Jamaicans went to Cuba to work on sugar plantations, and a small percentage went to the Dominican Republic and Puerto-Rico, where sugar production became important. At the end of the eighteen century, the British possessions in the Caribbean were Jamaica, Barbados, Antigua, Barbuda and Grenada, which had been taken from France in 1759 (Crusol 2007: 50). During the bicentennial of 'Wilberforce's Bill' on 25 March 1807, ratified by the British parliament to abolish the slave trade, a series of exhibits and events were organized in the United Kingdom, and in particular in Birmingham around the exhibit 'Equiano: From Enslavement to the Best-Selling Author' (Lord Morris of Handsworth 2007: 7). Emancipation touched the 'British world', meaning 'interactions between the British metropolis and the colonies, formal or informal', which expanded in several phases from 1815 to 1931, with a peak of the British Empire around 1921 (Vaughan et al. 2010: 4). After the general emancipation of British possessions in 1835, the evolution of migrations occurred in four main periods: from 1835 to 1885, from 1885 to 1920, from 1920 to 1940, and finally from 1940 to today (Marshall 1982: 8). Such a categorization solely based on old British possessions needs to include pertinent events in other Caribbean islands in order to offer a broader demographic, political and economic assessment of diverse migratory temporalities with the genesis and transformation of the societies in terms of differentiated strategies of development.

Putting circulation of ideas, skills and capital in homeland perspective

Migrations were once perceived negatively by both the countries of emigration and of immigration. Yet, the countries of origin and of arrival, just like migrants themselves, can profit from them, even if such advantages are difficult to quantify. One of the advantages for the host society is that immigration helps to deal with a qualified labour shortage. From the country of origin standpoint, one of the main advantages that is often cited is remittances. The Caribbean is a region marked by important flows of transfer, and it remains dependent on, and sensitive to, fluctuations in a globalized economy. The interdependence of its economy with the broader global one can be summarized in three factors: a complexity of family networks settled in the host society; a growing presence of direct foreign investments (IDE); and 'diasporised' labour. The combination of these three factors puts the economies of these countries in consistent need of foreign currencies; remittances to parents play a significant role in family revenues and, in the struggle against poverty, and they 'bring a certain degree of money circulation in all social classes' (Castor 2004: 159).

Paradoxically, the contribution of foreign currencies from the diaspora is of great significance: it represents more than 20 per cent of the gross domestic product (GDP) of some countries in the Caribbean and plays a crucial role in reducing the deficit in the balance of payments, while also limiting the consequences of potential natural catastrophes (Perchellet 2010: 155). For example, after hurricane Gilbert in 1988, Jamaica's GDP fell by around 2 per cent, even when a growth of 5 per cent had been forecast for the same year. Thanks to the contribution of remittances, the mobilization of Jamaican associations and collective actions abroad, the shock was partly absorbed. It is interesting to note, in some testimonies obtained in the course of the current research, the emergence of associations in Birmingham whose objective is essentially to respond to post-catastrophe emergencies in Jamaica. This is the case of the Jamaican association Erma Appeal, created in 1998 after a hurricane hit the parish of St Thomas, from which most of its members originate. Led by Alicia, it aimed to assist the population of the parish of Portland at a time of strong rains and devastating landslides:

> How to explain what I felt other than fear as I became victim of a fatal landslide? After my surgery and my rehabilitation in spite of my disability since this incident, I became obsessed with the idea of how to assist victims of natural catastrophe as I came back to Birmingham. I testified at a BBC radio programme and I sent an appeal for help that was heard so much that I quickly started the association two days later. My objective was to make the audience understand the necessity to react quickly and to structure the assistance to medical service of the parish of Portland

> of which I originate from. (Interview with Alicia, retired nurse, associative leader, Quinton, 20 November 2010)

Nevertheless, given that some of the money remitted returns to the host society for the family to purchase goods inside the country, 'the increasing dependency towards external revenue may discourage its beneficiaries from effectively contributing to economic activity' (Durand 1996: 423). Furthermore, the consumption of high technology goods or other imported goods unfortunately causes the departure of significant foreign currencies that have been transferred. Hence, although transfers are a means of reducing inequalities of revenues, the situation appears more ambiguous in some rural areas, where the majority of the beneficiaries of remittances are concentrated.

It is possible to observe a strong concentration of beneficiaries of remittances in Kingston, where there is an enormous unemployed population. In contrast with other parishes, the distribution of remittances is unequal and its impact on households varies in terms of levels of consumption and investments. Field investigation however, faces limitations in trying to bring more clarity to non-official sources of statistics on remittances.

My respondent at the Jamaican High Commissariat found that much effort is made by the Bank of Jamaica (BOJ) to achieve a better assessment of remittances, multiplying measures such as a higher rate of banking throughout parishes to block non-declared operators. If the question of the use of the human, social and financial capital of migrants raises concerns, it was in the 1990s that two approaches clearly started to emerge: control as a mean to contain the migratory phenomenon and development as a means to remove it (Guengant 1996: 107). In both cases, even when the objectives of control are met, development policies suffer from the brain drain.

> According to a United Nations report, 60 per cent of the most qualified Jamaicans live abroad. The question that I ask is the following: how many of these most qualified Jamaicans and which percentage it represents, are they ready to sacrifice and serve in order help the mother land? (Nettleford 2004)

Governments often perceived the migration of qualified workers as a 'brain drain' that depletes national human capital. This is exactly the case of Caribbean governments that have attempted to put in place policies to repatriate citizens abroad and their skills. Jamaica, Guyana, Haiti and Trinidad and Tobago are among the main exporters of intellectual capital to economies of OECD members (Ramphal Report 2011).

> 83 per cent of Haitian professionals left Haiti thirty to forty years ago. These people have studied here in universities financed by Haitian peasants' sweat. In ten years, we have lost 50 per cent of middle management professionals. Now, where are they? Never mind! They are often overqualified in sectors that are not connected to their academic background ... Today, such an erosion of our brains cruelly lacks in Haiti. What I say to Quebec: I know what the Haitian professionals contributed to the '*Révolution tranquille*' in the educational field, for example, because your city needed high quality French speakers. (Dr Chancy, secretary of state of animal production in Haiti, interview after the earthquake, January 2010)

The question here is whether the contribution of emigrants to national efforts at development can be appreciated in terms of knowledge and expertise transfer. If the issue of remittances is often treated independently of knowledge transfer, the two phenomena are related because the one who sends money often left years before as a student without an exact plan to settle or without wishing to deprive the country of a specific skill. How can we measure the impact of multiple and diverse mobilities on the homeland economy? The Jamaican nursing sector illustrates the extent of the consequences of the exodus of a qualified population and is symptomatic for the whole region. The recruitment of Caribbean nurses and professors from developed countries (the United States, Canada, and the United Kingdom) is key in the discussions between countries of the south and north. This phenomenon recently provoked a new debate that advocates a programme in line with a sound plan to stop the brain drain of the healthcare sector in particular. The loss for the sending country is greater than the cost invested in education of the staff that not only benefits the host society. Yet, agreements regarding the management of qualified human resources are often undermined due to the unequal political balance of power. The case in 1994 of a programme to recruit back professionals living abroad with the support of the Organisation of American States (OAS) is a good illustration of this (Gregoire 2011: 6). This situation reflects the profound imbalance between national policies and the host society, who benefit more, as they are willing to offer much higher salaries.

Box 2.1 The human capital haemorrhage in the Jamaican healthcare sector

Between 1962 and 1972, roughly 2,900 nurses may have migrated from Jamaica (Barret 1986: 39). A Caribbean Commission Health Report from July of 2005 underlines the negative impact that this has had in terms of budgetary education spending. Jane from the Jamaican Nurses Association (NAJ) explains the urgency of nurse migration and why the brain drain is an issue of concern. 'A positive interaction must necessarily be created between those of us who do not reside in Jamaica and those of us who have chosen to stay. Our thoughts and actions at the NAJ are geared towards Jamaica. This is why we think of ourselves as patriots in our vision and engagement to the development the nation.' Originally from the parish of Hanover, Jane's engagement with the NAJ is long-standing. She discussed the seriousness of the nurse exodus as the NAJ stays in touch with the situation (of nurses being directly recruited from Jamaica). 'The NAJ works with the Confederation of Black Ethnic Minority Nurses to address a broader Afro-Caribbean perspective. Poor working conditions and low pay continue to plague the Caribbean healthcare system as a whole'. This interview shows the difficulty of creating a sort of 'brain gain' to better utilize human resources by pushing for more 'circularity'. In the schemes for increased flexibility in work arrangements, this policy-level professional organization plays a positive role.

Source: Fieldwork, 8 November 2010, Milton Keynes.

Historically, the migration of qualified human resources of Haitian origin corresponds to two great moments tied to specific political junctures. The first moment refers to the period of the instalment of the Duvalier dictatorship during the 1960s. During the decolonization period of African and Asian Francophone countries, some Haitian professionals went into exile and some worked for education missions in Africa. This emigration of qualified human resources (RHC) is inscribed in an action led by Maurice Dartigue, the former Haitian minister of public education, agriculture and labour, who is well known for his reforms of the educational system and who became chief of the UNESCO mission to Congo in 1960. In that post, Dartigue created the National Pedagogical Institute in Kinshasa, which was approved in August 1961 (Kuyu 2006: 46). As a scholar in the United States, he obtained a master's in Rural Education in 1931 and worked to change the predominant approach: 'recitation, memorisation, repetition must be replaced by other elements of efficient pedagogy such as discussion, creativity and method, all skills acquired in the United States, as a system of analytical accounting for rural teaching' (Dartigue 1992: 22). In 1942–1943, he introduced Creole into school programmes and even created a Creole journal which published 5,000 copies. He sent the best staff to American and Francophone Canadian universities (Tardieu-Dehoux 1986). I met Eric during a conference on Haitian history and its contribution to Africa. He informed the audience about unjust complaints against the former education minister Dartigue, who once proposed to 'bring schools to the rural spaces where there is a need to value manual labour and the necessity for civic learning to make Haitians more conscious about themselves and the fate of their country'. Among Dartigue's reforms were progressive initiatives such as: the obligatory teaching of English; the creation of the Centre d'Art, a school initially devoted to spurring primitive painting production led by Dewitt Peters and the creation of the French Institute in December 1945, which was led by Pierre Mabille, and invited André Breton, who presented 'surrealism as a sparkle' (Bloncourt 2007: 24).

Box 2.2 Vivid memories of medicine as an ideal

Eric, a retired paediatrician in the Parisian *banlieue*, worked in Zaïre (the current RDC, formerly known as the Belgian Congo): 'Revisiting Haitian technical assistance in Congo means going back to a significant page of the history of the contribution to Congo that is unfortunately not well known. This is sad, because this emigration reflects the level of instruction that we had at the time in relation to others. Haiti witnessed the period of decolonization of countries as a member of great international institutions such as the Society of Nations (SDN), which became the United Nations.

Public representatives such as Edmond Sylvain, a former UN delegate to Haiti, and Maurice Dartigue, chief the UNESCO mission in Congo, worked very hard in countries that recognized our expertise in the repressive era of the Duvalier government. So, what can I say about my experience in Congo? This country will always be mine in a way after spending twenty-one years there, from 1969 to 1990. After finishing my medical studies in Port-au-Prince in 1960, I had difficult experiences in the rural city of Marchand-Dessalines, and this gave me a great deal of experience later in my post in Congo. I came to Congo as chief of the paediatric service, and I then became the dean of the medical faculty, which gave me the privilege of participating in the reform of teaching programmes. The last year of my presence as head of the department of community medicine made me proud to know that my social vision was recognized. What migrants bring and how others perceive us and how they see themselves is important, because no matter which level they attain, they know they have brought know-how and know-how-to-be, which are the most precious goods. I came to France and I continued to work, but in a different atmosphere. I left sincere and old friendships in my alma mater. It must be stressed that the new Congolese state only had twenty-nine university graduates. In Haiti, between 1950 and 1963, when I was left to specialize in Liverpool, England, there were approximately 500 physicians' graduates out of the medicine faculty. Out of that number, only 125 doctors remained to work in the country. This is the failure of the system. This explains my motivation to engage as a retired person who brought my expertise abroad: it is geared towards initiatives for an efficient transfer of knowledge in the homeland. As the *association Groupe d'appui médical* (GAMAH) reaches full speed, we do medical sessions and we give assistance to the *Maison d'Haïti* in the 20th arrondissement'.

Source: Interview with Eric, retired medical doctor, Paris, 10 May 2008.

The important movement of the emigration of many middle-class individuals, fleeing a politicization of higher education and frequent purges, significantly accelerated migration. Roger's statements as a medical doctor and also as a member of the *groupe d'appui médical* (GAMAH), reflect a sense of belonging that is nostalgic for a particular moment for his generation:

> I was in my first year of medical study in Port-au-Prince when a wave of repression hit one of my family members. I decided to leave for Quebec in 1965 to return later and work at my father's clinic. Sadly, once I finished my specialty in surgery in 1972, my family advised me to go to Montpellier and to continue my studies. It was a great university where about thirty of us resided. I even created an association of Haitians in Montpellier with three other doctors with the project to work in l' Artibonite, at Pont Sondé (Haiti). One of my friends came back and was jailed as a 'communist'. We discontinued this community health project. Once I became a doctor, I could not go back home; I decided to start a family and I got busy with other priorities. Today, I am retired and I am seriously thinking about a plan to return with my family to the department of Nippes where I am originally from. (Interview with Roger, medical doctor, Montpellier, 31 May 2008)

Collaboration between Cuba and Haiti came in the form of health centres (known as 'integral diagnostic centres') in order to contribute free medical assistance to millions of Haitians. Cuba has managed to conduct its unique cooperation with neighbouring islands and beyond in the Global South. The case of the Cuban medical presence in Haiti since the reestablishment of diplomatic ties (1995) points to a conception of geopolitics as a mutually beneficial interaction beyond the orthodox realist view of a zero-sum game. In reality, Cuba's role in Haiti has rarely been given the attention it deserves in terms of South-South cooperation, or what I term 'the cooperation of proximity'. Since 1998, Cuban doctors via health programmes have been able to reach the most remote rural towns in Haiti. Cuban medical diplomacy targets directly the most vulnerable sections of Haitian population. It is the majority who do not benefit from a serious national development plan who benefit most directly from such medical coverage. In a way, Cuban medical assistance also compensates for the brain drain of healthcare workers from Haiti fuelled by Western countries (the so-called 'friends of Haiti').

A second turning point in the emigration of qualified human resources came about at the end of the 1990s and the beginning of 2000. A political and socio-economic crisis hit Haiti during this period, when

Canada launched an aggressive incentivized recruitment policy. 'In 2000, 30,000 residency visas were delivered to the Canadian consulate in Haiti: accountants, nurses, computer technicians and university professors' (Gregoire 2011: 64). Globally, respondents and even less qualified ones said that, thanks to international migration, they acquired new skills and experience that could be beneficial for the country. Tommy is a baker residing in Vitry-sur-Seine:

> Initially, I discussed this issue of having so many well-qualified Haitians living abroad who could join me to share our knowledge. Every stay in Haiti, I tried to make progress on the project of my association, which I had created in 1994, the *Association d'aide au développement d'Haïti* (AADH) in the Parisian *banlieue* (94). Some friends discouraged me from doing this project, but I persevered. In April 2006, with the support of the managing committee and thirty members, we launched the construction of a building. One of my cousins, a retired agronomist who had returned to Fonds-des-Blancs, was a great help for my project of creating a bakery and a centre of technical training for young people. Hence, the idea was to teach them while they received a salary to start working on site. I also got precious help from contacts in *les Grands Moulins de France*. From Vitry-sur-Seine to Fonds-des-Blancs, it was possible to transmit know-how! These young people received a culture of entrepreneurship that gave them more autonomy and responsibility. It could be reproduced in other communes. (Interview with Tommy, associative leader of AADH, 8 May 2008)

In addition to these two great moments, however, the migration of low-skilled workers has never stopped. Non-qualified labour remains in demand for low salaries: this was the case on the island of Saint-Martin in 1985, due to French tax exemption causing a massive influx of foreigners, in particular for building hotels (Joint 2010: 179). The flow of international migrants increased rapidly, due to the opportunity for economic uplift and, social mobility it represents (Simon 2009: 93). Amos's journey reflects this hope: he left Gressier to become an electronic engineer in the city of Strasbourg in 2000:

> One of my cousins informed me about an important company that needed expertise, and I took the opportunity to leave Guadeloupe, where I was living a precarious life. Since, I got my son and my family to join me. I have participated in an old association where I socialize with good people; at least here there is a long settled Haitian

community. (Interview with Amos, electronic engineer and member of the association. Altitude, Strasbourg, 17 July 2010)

In a movement of migration from one country to the other, a two-way perspective on economic activity reveals the inextricable dynamics of three major actors: the migrant, the host country and the home country. The latter can encourage migration, because it is a source of foreign exchanges, or it can also discourage it in order to keep its vital human capital. These three factors can be influenced by the contribution of remittances, investments and public aid in the Haitian and Jamaican economy. 'A provider country could even decide not to take the option of migration if they did not accept it, while attempting to minimise the costs' (Bazin 2007: 12). In so doing, the path of Haiti and Jamaica in terms of development and the choice of migration policy offer some points of convergence. Chronic political instability, deficient institutions, apparently high rates of corruption, economic obstacles caused by emigration, a high unemployment rate and vulnerability to natural catastrophes are characteristics that are shared to some extent by both, even if significant differences deserve to be underscored.

The problem of qualified labour leaving Haiti and Jamaica is shared by both economies (along with others in the region), despite the significant external factors facing them having differed in various periods. One such example is the structural adjustment undertaken by Jamaica in the mid-1980s, which resulted in a major failure of efforts at liberalization and at opening up its economy. In contrast, the Haitian economy has stagnated over the past forty years. Apart from some brief periods of positive growth per capita in the 1970s and the end of 1990s, between 1961 and 2000 Haiti experienced a decline of 1 per cent a year in its real GDP per capita for a total decline of 45 per cent (ECLAC 2011: 7).

The multiplying effect of monetary remittances is most visible based on the five 'T's: 'transfer, telecommunications, tourism, trade and transport'. Indeed, the positive impact of the diaspora is essential for a number of sectors: exchange offices, banks, telecommunications, tourism (parish festivals, vacations, family visits), the exportation of ethnic food products and beverages (spices, rum, etc.) and air transportation. In my fieldwork, I found that the level of remittances sent home once a month is higher among Haitian migrants in France and its overseas departments than among Jamaicans.

A final aspect that deserves some attention deals with the informal economy, which is largely unaccounted for in the official statistics on the financial gains of the Haitian economy. Based on a report on employment, the informal economy and household consumption (EEICM) done in collaboration with the Haitian Institute of Statistics (ISHI), pertinent data was collected on the employment, consumption and buying power affecting the urban and rural population in Haiti. One component of the study concerns family consumption between 1999 and 2000, in particular for those located in urban districts, and it sheds light on the dependency caused by remittances in both rural and urban areas (Brisson 2003: 90). What is challenged here is the assisted behaviour that did not generate any production, as well as the level of consumption of imported products, which reinforced this state of dependency.

Box 2.3 The vicious circle of informal life

During an interview with the statistician Constance Torelli, who agreed to share some data collected in Haiti, the discussion revolved around the second part of this study regarding the contribution of the informal sector to the Haitian economy. For the beneficiaries living an informal life, the impact of remittances is much more positive than official reports reveal. Nevertheless, globally speaking, one needs to take a close look at the local level in order to get a better sense of the living conditions of so many households that are ignored by the state. Such households represent a big chunk in their contribution that unfortunately escapes the fiscal authorities. A more efficient approach could be put in place to allow the state to gain more resources from such an informal sector. Moreover, statistically speaking, much significant data is lost precisely due to the absence of a methodically sound data system to better trace households' contributions to the economy at large.

Source: Interview with C. Torelli, 26 June 2008, Institut pour le Développement (IRD), Paris.

Conclusion

> You know what the problem of the Caribbean is? Everyone came to do here what they could not have possibly done in Europe, and this attitude could not help but have historical consequences.
>
> — Gabriel García Márquez, Bohemia, Havana, 1979

The complex process of evaluating the true impact of capital flight, brain drain and uneven circulation of skills and ideas on the economies of countries that are structurally dependent on external resources continues to perplex. The arrival of remittances flowing to the Haitian rural areas after emigration led to the financing of many informal activities. Such is also the case in the urban milieu, because the informal Haitian economy is very widespread. Rural areas observed a proliferation of small street vendors, and markets are larger than in the cities. The circulation of American currency in the country contributes to the proliferation of stockbrokers throughout the main cities. This informal exchange currency market has negative effects on the statistics of the central bank, which cannot control the rates used by informal agents, nor account precisely for the sum in circulation within the territory. In closing, the Caribbean, characterized by small economies, continues to develop mechanisms of neo-colonial dependence. This can be clearly seen through the multiple remittances fuelled by intensive migration, which constitutes a constant fixture of Caribbean history and its culture of mobility. This trend is reinforced by the ongoing practices states use to provide labour in the region and beyond by implicitly encouraging it as a panacea or a safety valve to defuse social, economic and demographic pressures and tensions.

CHAPTER 3

Framing Jamaican 'community' in imperial landscapes: Citizenship, religious bonds and returns

This chapter is concerned with the ways that collective memory empowers Jamaicans residing throughout Great Britain to mobilize and attract human, social and financial capital. It explores how Caribbean families' receipt of remittances gradually moulds a 'community'. It examines the socio-spatial differences in the way the urban environment shapes the imaginary and consciousness of groups of people. Through the case of Jamaicans in the West Midlands, it comparatively assesses the notion of 'ethnic minority' as a social construct embedded in British proactive immigration and integration policy with a view to understanding how institutional imperatives favour some groups over others. In contrast with the case in France, an early top-level multiculturalism agenda seems central to understanding how political consciousness and associational activism became critical to creating differentiated citizenship.

Understanding memories reflected in the exhibition (Figure 3.1) helps us contextualize the Jamaican presence in a British society. Specifically, it casts light on the way a multinational state, in the post-Second World War context with a severely weakened right-wing of the political spectrum, chose to rethink the notion of 'subject' in relation to an awakening about the definitive settlement of immigrants out of the Commonwealth. Two visits to Birmingham contributed to my understanding of the complex relationship between memory and postcolonial discourse framed in the discussion of migration issues in Great Britain. The first visit was to the exhibition at the Birmingham Centre in September 2010; and the second was in October of the same year to meet with a civil servant working for the Equality & Diversity Commission (previously known as the Commission for Racial Equality (CRE). The exhibition was of great interest in revisiting the framing of the city's history, and it reflected indirectly on my questions about the contrast between the immigration policies of France and the United Kingdom. Indeed, both countries took distinct routes after the Second World War.

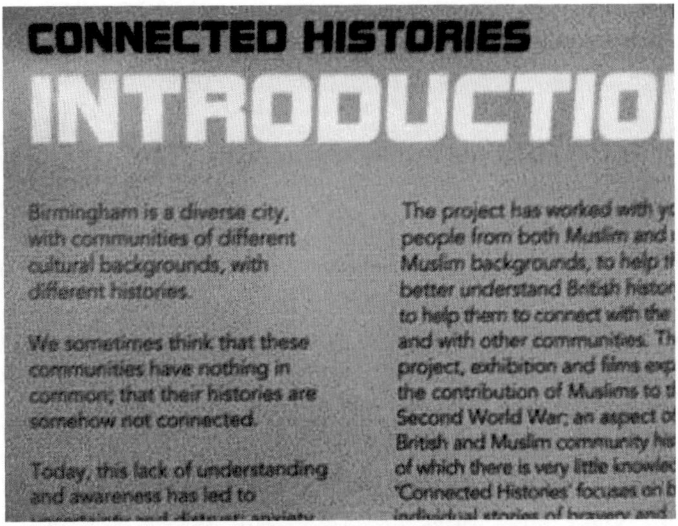

Figure 3.1 Exhibit 'Connected Histories'.
Photo by author: Birmingham Centre (September 2010).

The economic boom of the post-war reconstruction of Britain and France drew in numerous immigrant workers as a response to labour shortages. In Britain, the first substantial group of immigrants came from the Caribbean, and their pattern of settlement was determined by the employment structure. Where Portes and Rumbaut (1996: 84) focus on three main areas – host government policies, working conditions and specific characteristics of communities – I demonstrate the pertinence of sense of belonging expressed through distance and proximity in this population towards the host society. If numerous families are beneficiaries of remittances from the intensification and diversification of migratory flows, it is now necessary to interrogate the emergence of the Jamaican community in a detailed manner. It is worth examining how certain migrants who have already settled seek to regroup their efforts to create a 'community', and this question has been debated among them.

My interview with Saptal Hira, team leader of the Equality & Diversity Commission at Birmingham City Council, reflects the general British approach towards immigration policies. He clearly explains the role of a structure aimed primarily at monitoring the degree of importance given to ethnic minority issues and the ultimate goal of community cohesion:

> One of the main points of the policy of community cohesion is to give the City of Birmingham the capacity to manage with fairness all the segments of the population's needs. It is an integrated mechanism that reconciles action plans and allocated resources. It is based on four aspects: to live together in harmony; to share an interest in solidarity; to maintain ties and relations; to be proud of the city. A set of strategies and action plans forms the basis of our framework. The key points target violence and extremism prevention, and in particular the youth. We anticipate a project to give them education with immediate job opportunities. An electronic model is in place to keep track of religious organizations that support our actions through a web of spiritual leaders who are legitimate in the eyes of different communities. In this sense, we work for example with a group of representatives of the Afro-Caribbean community by the name of United Building Legacy (UBL). (Interview with Saptal Hira, civil servant, Birmingham City Council, 25 October 2010)

Contrasting the French and British societies via a historical consideration of their respective Caribbean populations entails a questioning of two dissimilar and common contexts. My concern is not to reinforce the specificities of each case, but rather to go beyond the national milieu (French Republicanism versus British Race Relations) that is so frequently the focus of the discussion. I intend to demonstrate why and how the colonial past continues to play a salient role in the reality still alive within Caribbean societies. Such an approach allows us to better appreciate the experience of the two populations studied here. Before proceeding any further, I provide context for my analysis and interpretation of how these colonial ties have evolved. The questions to be answered are: How have relationships evolved, and what impact do French and British immigration policies have on the modes of settlement of these populations? The latter question is explored with specific reference to their transformation (or failure to transform) from 'immigrants' to 'communities', and the extent to which they are perceived and treated as 'ethnic minorities'.

Pluri-belonging and citizenship: The challenge of multiculturalism

I begin by clarifying the notion that multiculturalism is opposed to a stricter notion of citizenship '*à la française*'. Indeed, multiculturalism simply does not mean what most of its critics think. This section raises an ongoing and controversial debate on divergent views in political theory about multiculturalism and minority rights within a liberal context. 'Multiculturalism understood as the celebration of difference and the relativism behind it raises fundamental challenges to the quest to understand others' (Fay, 1996). This refers to something crucial in the contemporary globalized world: that people significantly different from one another are in contact with, and must deal with, each other. As such, it insists on cultural and social integrity and on esteeming this integrity in others.

The term 'multiculturalism' functions as an umbrella that covers many different ideas, and there is no consensus among political philosophers about the meaning of the term. However, the common denominator that seems to bind different multiculturalists is unease about the way mainstream liberal thinking treats the issues of diversity and pluralism. Canadian philosophers such as Will Kymlicka and Joseph Raz justify granting collective rights to Canadian Aboriginals to ensure that they can sustain their cultural way of life with regard to economic interest, such as power, respect, or social status (2002: 367). Such a concept does not mean, and has never meant, different cultural and ethnic groups being left alone to get on with whatever activities they choose; quite the opposite. Indeed, Canada has done most to define and elaborate the concept, since it is quintessentially an immigrant society. Policy-making stresses active dialogue between cultural groups, active attempts at creating community cohesion, and the acceptance of an overarching Canadian identity.

Arguably, multiculturalism has always been a more sophisticated process in practice than its opponents usually recognize or indeed its supporters. The fact that immigrant communities do gradually absorb the core values, and even the underlying cultural myths of the host society (for example, the phrase '*nous sommes tous gaulois*' or 'we are all Gallic' highlights the French attitude towards overseas populations) is testimony to an emerging multicultural credo whereby shared values are more openly insisted upon. Recently a debate between two contenders for leadership of the right-wing French republican party, Les Républicains (LR), underscored the tensions between two Frances: one proud to recognize its diverse population given its colonial past; and one restricting its identity to an exclusive belonging rooted in the Hexagon with no multicultural branches.

If the many nations that have adopted multiculturalism, such as Canada, Britain, the United States, and others, have experienced both challenges to and support for multiculturalism from diverse groups, one must recognize the complexity of multiple socio-economic issues emerging out of such policy. In nations such as Switzerland, Belgium and Spain, liberal democracy coexists with multiculturalism and questions arise concerning the relationship between these two main principles. How does multiculturalism fit into liberal democratic society? Does the principle of multiculturalism support liberal democracy or undermine it? More specifically, what aspects of multiculturalism affect liberal democracy and how?

I take as a point of departure that the challenge of multiculturalism is primarily a challenge for liberal democracies or for societies that are seeking to democratize their public sphere, that is, to invite and increase citizen participation in civil society. It should therefore come as no surprise that it is in societies with cultural diversity and a high level of commitment to democracy that debates, contestations and movements surrounding multiculturalism have been the loudest (Tiryakian 2003: 20–39). Why? It is certainly because it is in the nature of liberal democracies to engage or encourage contestations in the public sphere. It is also due to perceptions that the state will make resources available in response to certain grievances being aired in the political process. This is often the case in the French context, where public demonstrations of discontent and strikes are common practices in all sectors of the society. In fact, among the inherent tensions between multiculturalism and liberal democracy, the most problematic remains the conflict between equality of political rights (fundamental to liberal democracies, and France in particular), and the notion of collective rights (fundamental to multiculturalism, especially in the British context). Such tension can be overcome, and I argue that multiculturalism in fact supports certain elements of liberal democracy such as political participation. The case of French Guiana, and Haitian-born citizens and non-citizens' political engagement in the electoral process of 2010, will be discussed further in a later chapter.

There are those who question both the demand for special group rights or minority rights and for multiculturalism. Huntington, an opponent of the American multicultural theory, states that 'proponents of multiculturalism wish to create a country not belonging to any civilisation and lacking a cultural core. History, they would argue, shows that no country so constituted can endure for long as a coherent society' (1996: 305–309). My argument is that history shows precisely the opposite. Throughout the world, minority groups have become increasingly assertive, demanding recognition and rights, while the powerful, seeking to protect their own positions, have responded with repression and violence. In a pluralistic society, all groups should accept the need for scrutiny from others, as it is the condition of producing mutual respect, rather than undermining it (Erdem 2002: 8–10). The argument that the spread of pop culture and consumer goods around the world represents the triumph of Western civilization denigrates the values of other cultures, while trivializing Western culture by identifying it with fatty foods and fizzy drinks. The essence of Western culture is the Magna Carta, not the Magna Mac. That there are significant differences between modern and traditional cultures is beyond dispute. If numerous families are beneficiaries of remittances from the intensification and diversification of migratory flows, it is worth examining how certain migrants who have already settled seek to regroup their efforts to create a 'community'. To do this, I shall examine the emergence of a Jamaican 'community' in Britain, which is bound by vivid collective memory and a sense of pluri-belonging and expressed in its socio-spatially differentiated presence.

Resisting and surviving in the West Midlands

Don't let them fool ya ... We've got a mind of our own ...
— Bob Marley, 1980

The West Midlands contextualizes the Jamaican presence through socio-spatial differentiation caused by the way the urban environment shapes the imaginary and consciousness of populations. The way space and memories are intertwined in the storytelling of some of my respondents' challenges visibility and activism as an aspect of citizenship. Birmingham lies at the heart of the second largest urban agglomeration in the United Kingdom and the city's minority population continues to grow, mainly due to a relatively high birth rate in a youthful population (Owen 2001: 58). It has a very powerful local body, the Birmingham City Council, the largest employer in the West Midlands. One key piece of legislation, the 1962 Commonwealth Immigrant Act, caused a remarkable surge in Caribbean migrants. The Birmingham City council is a very powerful local body and represents the largest employer in the West Midlands.

The first waves of Jamaicans migrating to the United Kingdom are symbolically remembered by their arrival aboard the ship SS *Empire Windrush* at Tilbury in the south of England in June 1948. Nevertheless, before this event, 'marines came from the Caribbean, others worked as technicians in the arm industry in Liverpool' (Thomas-Hope 1995: 119). Shawn, a retired army man, arrived in 1944 at Liverpool and then settled in Soho, Birmingham. At the time of interview, he is an active member of the Windrush Foundation, as he wishes to keep the memory alive for the younger generation:

> During the summer of 1944, there were about one thousand Caribbean marines employed. I left behind my mother and my sister in the Jamaican parish of Westmorland. I received a very intense training in the medical sector and up to then everything was fine. I went to live with my cousin settled in Liverpool and, there, I realized that the post-war climate was quite complicated for us. It is like they wanted to discourage us from settling in their neighbourhoods. The shock was profound on me: I could remember my mother telling us that we came from a Scottish ancestry who came as adventurer in Jamaica! Anyway, in June 2008, the Windrush Foundation

> managed with the support of the New Church Testament of God NCTG to raise awareness about the fragile living conditions of some of those elders in renting a flat or in using the NHS. (Interview with Shawn, retired, Soho, 19 October 2010)

Thus, scattered over London or Birmingham, this community of Jamaican settlements expands as far as Truro and Liskeard in the west and, as far east as Norfolk. At the beginning of my fieldwork, I did a pre-survey (24 to 30 June of 2010) with institutional actors such as officials of the High Commissariat of Jamaica. The idea was to gain a broad sense of where to locate Jamaican associations, particularly in London and Birmingham, and to get a sense of the institutional structure of the Jamaican diaspora throughout the United Kingdom known as JDUK.

The new migrants suffered from unsanitary and overcrowded housing, and from a lack of institutional support to succeed in school and to access health services. Kwas, a militant of the oldest Jamaican organization, whose goal was primarily the struggle against discrimination, consented to revisit his painful experience of settling in Britain with his family in 1959 in Handsworth:

> My parents were encouraged to come to the 'motherland' after the Second World War to work towards reconstruction. They awoke from an atrocious dream where they felt humanely diminished. There was like a cultural shock in me even as a child who could not understand mockeries from little white schoolmates. Later my path was set to launch an association to fight for equality and respect of rights. (Interview with Kwas, member of the African Caribbean Self Help Organisation (ACSHO), Handsworth, 4 November 2010)

Through a meeting with the United Building Legacy (UBL) association, I became involved in some educational activities with the Caribbean community. Simone, born in the parish of St Ann, is an educator based in Lozells since 1988. She discusses some of the problems with the educational system, which needs a 'different approach' in tune with the Afro-Caribbean community. Our discussion sheds light on some paradoxes of the legislation and the perception of some in British society:

> I am an 'Afro-Caribbean' because this category reflects my heritage from Africa. I cannot think of a better term to express my feeling. Growing up, I used to notice Black

> people with a Caribbean accent. Today, it is less the case due to the greater presence of people coming from the continent of Africa. At the same time, white individuals could be viewed as British even when they are Europeans since the reunification of Germany. Demography has changed the notion of Afro-Caribbean, as it allows us to recognize a broader diversity of people out of a migration tied to colonial history, whether from the Caribbean or India, from Pakistan or other African countries. What creates problems and ambiguity is creating laws that do not necessarily change the societal perception of us. (Interview with Simone, an educator and member of the UBL association, 6 November 2010)

This discrepancy between legislation and policy towards a 'minority' and the perception of those on the receiving end reveals a similar result in terms of common practices in such different countries as France and the United Kingdom, sites of trade and the slave trade in their common colonial pasts. In that sense, the process of decolonial thinking is central to articulating any informed analysis of Caribbean migration and the strong link between (under)-development and the settlement of communities throughout European cities. The discussion shows how a state constructs category to designate a segment of its population when this population may already be granted citizenship. A comparative reading of the immigration legislation of the French and British institutional frameworks from 1871 to 2012 clarifies the ways in which this has been done.

Whether it was about seeking asylum or done in a clandestine fashion, the journey to Britain often started by means of a family member or contact person. In almost all the cases interviewed twenty-eight family reunifications was the initial cause of departure. Only two of them came exclusively to pursue their university studies in the beginning of the 1990s while another had an employment contract. For this form of essentially familial immigration, the weight of family networks is decisive in the relationship of immigrants to a British society that perceives multiculturalism as recognition of tolerance and affirmation of difference. For those who feel Jamaican, a sense of pride in this identity seems tied to a form of profound loyalty nourished by the memory of belonging to the Jamaican nation:

> I am proud and passionate about my homeland and I think that most people of Jamaican origin share my view about living here in the UK. When it comes to support Jamaica, the diaspora is always present, even if it is not always sure where the

money goes. This does not stop the initiative to show solidarity by our engagement as a community. (Interview with Anna, a militant activist born in the parish of St Thomas and settled since 1961 in Birmingham with her parents, 28 June 2010)

One of the characteristics of the Jamaican family that became prominent in my research is a constant parental absence during childhood. This trait reveals how migration affects a sense of belonging for some of my respondents. Jocelyn, a sociologist, looks back at her chaotic adolescence in the parish of St Thomas when she was separated from her parents, who were living in Birmingham:

> Honestly, there is a gentle alienation here brushing off African-Caribbean people in England. An immigrant child fights more fiercely than an Asian kid, who gets more family support. In my case, I felt like a stranger coming to this household separated so long from my mother. She did not know me and she had a new family here. It took twenty-five years to begin a new relationship and to familiarize myself with this new home. Being a sociologist, there is an understanding of the whole migrant labour system and I relate the stories of so many mates who just like me were treated badly as outsiders once reunited in England. This is the dilemma of the first generation who arrived much too late to be with their parents already settled here in the 1950s-1960s. Home is a psychological feeling. Do I feel British? Good God no! I played my part in the society and I am very conscious being a citizen, taking responsibility; I even was involved in politics in the county of Perry. However, I cannot as a conscious African woman having the memory I got of my people, the maroons from the parish of Portland, want to belong to somewhere people actually do not want me to belong. I mean to say that UK is a multicultural place that respects cultural difference unlike the assimilation imposed by France. I visited France and I could see the colonial atmosphere. Coming from Haiti I know you understand my point because your country has a long history that many ignore right here! (Interview with Jocelyn, sociologist, Handsworth, 22 November 2010)

Five of respondents identify as feeling both Jamaican and British. Pastor Vlad, a theologian and journalist at BBC Radio, points to the following:

> I do not see any conflict of being simultaneously Jamaican, a real sense of 'Jamaicanness', while assuming my roots in the Parish of Sainte Catherine and fully feeling British. I've lived more than forty years, longer than in my birthplace. So, I am made of a British part since I have lived here for so long, I continue to contribute to the society, but this should not make me deny a part of me tied to the Jamaican history. (Interview with Vlad, theologian, Aston, 25 November 2010)

Lee is one of the few respondents to put his British belonging first:

> I do not share the view of those who said that they are not British because they do not feel loved or because the British society does not want them. Who are them to tell me who I am? I have always felt with a Christian consciousness that transcends all prejudices and self-withdrawal. I feel that we have inherited a British culture that has given us great institution and above all Anglican values that fits me well. I recently had the honour of receiving the title of Member of British Empire (MBE). (Interview with Lee, athletic coach, Handsworth, 25 November 2010)

Others, who feel Afro-Caribbean, think well of the political discourse spread after the census which officialized this ethnic category. Others such as Mussa, a Rastafarian artist born in 1956 in Birmingham from a Jamaican family, or Kwas, a Pan-African militant who leads the oldest Jamaican association in Birmingham, share a position that echoes the Pan-African ideology:

> This sort of brainwash that causes a rift between us and Africa and our belief of Great Britain as our motherland with a governor general representing the Queen in the country that just about to celebrate its fiftieth independence anniversary in 2012 does not make sense. Such hostile environment helps reinforce a common identity among Caribbean and African, hence the reason for my association title African-Caribbean since its inception in Handsworth. (Interview with Kwas, militant Pan-African, Handsworth, 4 November 2010)

The migrant individual is invariably compelled to adapt to all changes, often tragic ones, throughout their experience of migration (Table 3.1). Some migrants become illegal after visa expiration. It has been estimated that approximately 30 per cent of Jamaicans living in the UK are without papers (Martin-Johnson 2009: 222). These individuals are incapable of proving that they are legal British citizens and at the same time do not want to return 'home', to Jamaica. They doubt the efficiency of the voluntary return programme designed by Jamaica for illegal immigrants (AVRIM). Others are deportees, numbers of which to Jamaica have been increasing since 1997. Thus, some immigrants are tempted by the opportunity to return home while others, as they acquire new citizenship, use local religious networks with shared goals that transgress boundaries to renew with deep spirituality.

Table 3.1 Sense of belonging to the Jamaican nation

Sense of Belonging	Jamaican	Jamaican/ British	British	African-Caribbean	Total respondents
	11	5	2	10	28

Bonding through religious networks: Pentecostal spirit of memory

> Singing the Lord's song in a strange land is no easy business nor is it a song best sung alone.
> — Dr Joe Aldred, 2012

Although there is abundant literature on certain transnational ties, such as remittances and hometown associations, little has been published on migrants' transnational religious activities. This is noted by Levitt (2003: 847) who claims that 'while migration scholars now generally acknowledge the salience of migrants' economic, social, and political transnational activities, we have largely overlooked the ways in which religious identities and practices also enable migrants to sustain memberships in multiple locations'. I view religion as what Levitt called 'a contingent clustering of diverse elements that come together within to-be-determined spaces riddled by power and interests'. As a result, actors and ideas travel at different rates across different levels and scopes of the social fields where they are embedded. Yet, we still assume that religious practices subscribe to a particular territory when evidence-based field research often reports otherwise. Rather than assuming that religious life remains primarily within contained spaces, I start from an assumption of circulation and linkages. The connection has to do with the way congregations' diasporas form and become alternative spaces of belonging and reference points for establishing collective identities in a society that constructs them as the 'other'. Belonging is a method of remembrance; it entails constructing a collective memory from the natural, surrounding environment (Lovell 1998: 14). As many relations are mobilized, this network functions by anchoring itself in the territory. The spiritual belonging is marked by sacred acts, which are resources that are capable of uniting individuals. Such a transnational dynamic of vibrant space contains actions that transcend the institutional settings. Understanding the nature of the religious community to which one belongs requires a prior understanding of the significance of Pentecostalism and members' participation in this form of religious expression.

Historically, the Pentecostal Church of England comes from a division between Quakers, Methodists, and Baptists who were dissatisfied with the Church of England, which was itself born out of a split from the Catholic Church. The Pentecostal branch reached the Caribbean and, in particular, the island of Jamaica around 1880, passing through the Southern United States of America (Cleveland, Tennessee). This church would go on to experience some changes with the arrival of hundreds of Caribbean individuals after the First World War (Edwards 1998: 13). This Pentecostal branch would emerge in 1923; it modified its ritual by adapting it to the Caribbean environment. Two rival branches now settled in Birmingham: The New Testament Church of God (NTCG) and the Church of God of Prophecy (CGOP). At the end of the mass, I met with Nick, a member of the NTCG:

> Our church (NTCG) organizes very cheerful celebrations animated by a young orchestra and this gives an appropriate ambiance for human warmth every Sunday. Here, in a harmonious ambiance, all generations get together and this is a place of privileged exchanges. I grew up in this Pentecostal tradition and my parents, from the parish of St Ann, lived in this ambiance as well. (Interview with Nick, born in Birmingham, in charge of the cultural affairs of NTCG, 27 June 2010)

The NTCG, located at the heart of the district of Lozells (Birmingham), is primarily occupied by Jamaicans and also individuals of Caribbean origin. In the company of a joyful audience and many families, this space serves as a fraternal grouping. The space is experienced as a moment of reconnection among people of the same community. The relationship to space and the significance of the Pentecostal cult begs the question of how religious belonging may or may not have an influence on associative engagement within the communities of Jamaican origin. Some of our respondents talked about their religious experiences, insisting on the solidarity and the spirit of communion that they feel throughout the Pentecostal services in contrast to the British Anglican church rituals.

Nick's and Dean's stories underline the evolution of Pentecostalism in a community faced with social deprivation and the harsh insertion into British society in the course of their settlement. They still remember tough situations where they felt embarrassed and humiliated, such as when they were attending a British Pentecostal mass in a hostile atmosphere. Even when it is difficult to fully understand, the influence of religion on individuals' migratory experience (departing from or returning to the homeland) makes it appear as if the migrants were travelling with their God (Levitt 2004).

> We had a small Pentecostal church, but also a Wesleyan Methodist mission in my native village. The Church of God of Prophecy, which originated from the South of the United States of America, organized missions around the world, and this is how it became rooted in Jamaica and across the Caribbean. Thus, I grew up with this belief without asking myself many questions. (Interview with Dr Vlad, theologian, Aston, 25 November 2010)

Nancy, a native of the Jamaican parish of Clarendon, has presided over the Council of Black Led Churches (CBLC) since its creation in 1988 after the riots in Birmingham. She underlines its prominent role, composed as it is of approximately 200 Black Christian churches (Baptists and Pentecostal), and their relationship with the local authorities:

> The CBLC and the Pentecostal Church have studied the culture of gangs at Birmingham and the theology of reggae in order to explore the future of British Caribbean Pentecostalism in a society where the notion of white supremacy, even at a faith level, remains widespread. CBLC defends the prophetic role of the traditional Black church as an inherent part of its mission. Our interventions invite individuals to reflect on the contribution of the new generation of academics and Afro-Caribbean theologians, whose works spark debates. (Interview with Nancy, social worker, Hampton Row, 7 December 2010)

For Dr Vlad, a native of the Jamaican parish of St Elizabeth and an ecumenical theologian at the Church Together in England, the real weight of the Jamaican Pentecostal Church settled in the United Kingdom. The City of Birmingham places great significance on the strength of its networks. Three elements contribute to its key role within the Jamaican community and beyond: spiritual memory and social capital, institutional structure, and socio-economic resources:

> I feel that the Jamaican church has had a significant role in the diaspora, which may be profitable for Jamaica. Although the Catholic Church was for so long the main denomination on the island, it is only when the Church of England came into existence that the majority of Caribbean Catholics came to seek something else. There was a stereotype, which persisted, which is that Jamaica is different from the other islands. In reality, the Pentecostals are just a part of the equation, because it is the Anglican Church that dominates and then the Baptists. In the time of the colony, there was no Pentecostal church but many Baptists. A pastor by the name of George Liele, a Native Baptist Church, came from America to settle in Jamaica and created the Baptist movement. This is the start of a missionary Christian history in Jamaica that would mark a whole period of spiritual agitation, at a time where prayer meetings took place all over the island. This history attests to the weight of the Church as an institution in the lives of the Jamaican people whether abroad or at home. This religious bond gives a sense of continuum as a ritual of remembrance. This structure guides him in its tribulations. The 'Black Church' is a myth spread here in England that reinforces the place that the Church occupies as a space of power. (Interview, Aston, 25 November 2010)

Embodying returns narratives: 'Exodus ... Moving out the people!'

Transnational studies of migration have often examined the experience of return and underscored the complexity of collective memory in this process, which is full of ambivalence, from the perspective of the migrant. Sometimes described as a circular movement based on economic survival, or even treated as a 'double passage after a long presence abroad' (Thomas-Hope 1985; Conway 1992), return continues to feed the migrant's memory and imagination. So, a category of migrants can be considered as in transit while working towards a better economic situation with the conviction of returning home one day. Kate, an educator and a native of the Jamaican parish of Manchester, is fifty-eight years old. She remembered her difficult settlement with her family in the East Midlands:

> My parents came at the start of the 1960s from the parish of Manchester. Several families settled in the City of Derby in the East Midlands knew each other back in Jamaica. This gave us a small community proud of its Jamaican identity and with the hope of being just in transit as economic migrants. Once retired, my parents went back to Jamaica. They were quite lucky to attain their goal: to raise us and then to enjoy a quiet retreat in the homeland. (Interview with Kate, educator, Derby, East Midlands, 10 December 2010)

The return visit concerns another category of return migration that is quite interesting. Indeed, it is about frequent visits under the forms of entrees and departures of 'tourists' and business people from abroad. From the point of view of the origin society, this type of movement occasions a number of benefits for the local economy and the tourist sector. Hence, some states develop strategies to attract members of the younger generations and the retired. Research (Thomas-Hope 1981; Nutter 1989; Byron 2000; Plaza 2002) underlines the phenomenon of young generations undertaking frequent trips back and forth. Other young returnees continue their economic activity or prefer to invest in the tourist sector. These are entrepreneurs who are interested in real estate as well as the telecommunications sector. This reflects a significant trend of return migration among young Caribbean British citizens. Dean, one of my interviewees, who participated in the Jamaican Diaspora Convention at Ocho Rios in June 2011, explains:

> I am part of the delegation 'Diaspora Future Leaders' and this is my first time to participate at this convention. I introduced my society of new technology and I did networking with some contacts and with the JAMPRO (Jamaica Trade & Investment) representative. My goal is to return next summer to probably settle in the parish of St Ann. I remain optimist and prudent in my initiative. Anyway, I have to take risk otherwise I may waste my small capital. The culture is a bit different here from Birmingham; however, I feel home and even the obligation to succeed in my parents' homeland. We'll see! (Interview with Dean, young entrepreneur, London, 9 November 2010)

In this case, it is a returnee born of Jamaican parents in Birmingham who is thinking about succeeding in his return to Jamaica as an entrepreneur. He is a qualified professional. I discussed with the representative of the governmental Jamaican agency JAMPRO in London, how the young generation is able to reconnect with Jamaica:

> There is a desire to reconnect with the homeland. I was told in a barber shop about a conversation with a young man of Jamaican origin that appeared very 'British', to whom they asked where he came from. This shows that one never is truly accepted as fully part of the British nation in spite of efforts that could be made to do so there is the vague perception to not belong if one does not physically resemble a 'white Anglo-Saxon' protestant, and so you are evidently an outsider! Then the young man at the barber shop asked: does the country still need us as non-Europeans? I conclude from this tale that the Jamaican High Commissariat at the trade sector has the ability to initiate creative promotion of the culture and entrepreneurial potential towards the younger Jamaican generation settled in the United Kingdom. One of our initiatives in preparation to the next Jamaican diaspora conference is to attract young people with an interesting profile to expand the entrepreneurial landscape of the country. This could be a win-win situation! (Interview with Lawrence Jones, commercial representative at JAMPRO, London, 29 June 2010)

In an exchange with research assistant Dr Natasha Mortley, at the University of the West Indies (UWI), I learned about a field survey exploring the behaviour of the young immigrant youth in the tourist sector during their sojourns in Jamaica.

> There is a generational distinction to be made: the pioneer generation who participates more to the homecomings and to community-based activities within the parish, while much younger generations stay more in hotels that represent a direct contribution to tourism sector. This youth are motivated by the desire to discover an ancestral land, a way to revive their memory and to reclaim their roots. They participate in weekend parties at Negril during summer or at carnivals before Easter. Their behaviour varies depending on the generation and it is difficult to come up with a clear typology generation-expense. (Interview with Dr Natasha Mortley, Mona Campus, Kingston, 8 June 2011)

The returnees of a new generation of Jamaicans in Great Britain do not always match the profile of older returnees. These individuals are often qualified professionals with a project to work on once back in the Caribbean. Their return may be a continuation of their professional career, as is the case for Leo who created an Association for Returning Residents (ARRR). Its aim is to foster a sort of idyllic reconnection to the motherland. For some returnees who are victims of discrimination and class politics, the re-adaptation in Jamaica becomes difficult. As such, Leo's statement supports the conclusions of studies about the obstacles and difficulties for the successful reintegration of returnees into their societies of origin (Plaza & Henry 2009: 16).

If the goal in return migration is often to return to a situation of emotional security and a level of social comfort at home, new questions and problems emerge for families and governments who encourage this practice. One of the difficulties noted during my interviews in Jamaica was the rising price of housing due to higher levels of returnees and expenses for local municipalities. Certain parishes are trying to think of attractive programmes to assimilate returnees as tourists. The mayor of Parish St Ann welcomed the Jamaican diaspora at the Convention in 2011:

> The movement of the diaspora is greater than all of us, greater than any institution and for us an urge to appreciate its true potential. This should be a constant force in order to make the local economic sector more dynamic. (Statements from Mayor of St Ann, Ocho Rios, 16 June 2011)

One of the participants at the convention, Kate, commented on the issue of Jamaican returnees and detainees in the United Kingdom and the best practices for better reintegration into Jamaican society.

> The time has come for our delegation to tackle a delicate question about national security and criminality. We think that Jamaica will need our assistance and expertise on these issues because we are also victims of the negative and often degrading image of the country as displayed in the media around the world. (Kate, university professor, Ocho Rios, Parish St Ann, 17 June 2011)

In the 1990s Jamaica received a growing number of return requests from citizens experiencing difficulties abroad, beneficiaries' preferential customs fees, or other demands to facilitate their repatriation. The number of returnees from the United Kingdom alone rose to about a thousand in 1994. An office was established at the Ministry of foreign affairs to process returning residents' files and customs for vehicles and other goods. Criticisms of the relatively limited role played by this structure led to the creation of a small department for Jamaicans abroad compared to the rest of the ministry. Later, a programme for returning residents was created in 1993 to help Jamaican citizens with job and investment opportunities. The example of Leo, a native of the parish of St Thomas whom I met in Kingston (June 2011) illustrates the difficulties some have in returning.

> It all started at Pegasus Hotel when I returned to Jamaica at the end of 1977. In five years of existence, my association of Resettlement of returnees (ARRR) expands its action in all parishes in assisting about 65 per cent of 51,000 returnees. (Interview with Leo, retired and associative militant, Kingston, 21 June 2011)

While aware of the problems returnees are facing, Leo believes that the services offered are moving in the right direction to ease their adaptation. His work is not always 'understood' by Jamaican officials, even when it seems 'pertinent' to those who work on the returnee issues (confirmed by Professor Thomas-Hope, UWI, 8 June 2011). Leo explains how he observes with anguish some delicate situations for returnees, in particular those from the United Kingdom.

> When they return, some are identified to the foreign place where they resided. For example, they are mocked by calling them 'British' or other labels based on their different accent or other acquired traits abroad. The image of the returnee is often assimilated to eccentric behaviours or clothing excess in contrast to the people who never left. They often feel misunderstood and astonished to be identified to a foreign country where they certainly lived but not necessarily assimilated to. So, it is difficult for them to realize the difficulty they experience to reinsert themselves in the supposedly home society. This is part of a broader dilemma and ambivalence of the Caribbean culture and its own contradictions. (Interview with Leo, retired and president of the association ARRR, Kingston, 21 June 2011)

Box 3.1 Reconnecting with the homeland

To return to Jamaica without bringing much from working abroad, or without showing signs of success, is often considered an enormous humiliation which causes people not to return at all or even delay their departure. Leo tries, through his association ARRR', to provide people the best path to reconnect with the homeland. 'Some associations do not receive any funds from the Jamaican government, who tends to manipulate associations like ours to better legitimate their political communication for investment in the parishes. From a local politics point, the relationship between state/diaspora depends on the diaspora structure itself'. ARRR's services gradually expanded to ensure a smooth reinsertion for returnees, in particular those coming from the United Kingdom, some from Canada and the United States. Among the services offered include counselling, real estate, banking services, pensions, nursing home facilities and services for mentally challenged and disabled. ARRR's objective is to advocate on behalf of returning residents. One of its services is assisting returnees who have purchased properties without the assistance of an attorney and also campaigning to protect returning residents from becoming victims of fraud.

Source: Interview with Leo, retired, associative member, Kingston, 21 June 2011.

Conclusion

The symbolism of return helps locate memory through the physical presence of home. What has been less well understood is how in opposition to the possibility of returning, the migrant may ask the following: 'to stay or to leave can contribute to development, to leave to better return or to stay in migration to allow other to live?' (Charef and Gonin 2005: 11).

More broadly, we can conclude that the longer migrants remain abroad, the more complicated the process of reinsertion in the country of origin becomes upon return. Yet, this discussion points to a production of collective memory around the theme of returning that does not stop the affirmation of a legitimate right to belong to a nation of origin. In order to work against what they perceive as a misinterpretation of their identity, some returnees tell many tales describing their acts of courage and 'bravery' during the migration experience, such as being exposed to humiliations and inhospitality. This is where one of the ambiguities of the relation between migration and development remains, whether from the standpoint of the returnee or the origin state that seeks to capitalize on migration gains. The state does not necessarily take sufficient measures to help those who wish to return. On the other hand, those who have left do not necessarily cease wanting to go back.

CHAPTER 4

Beyond French Republicanism? Locating in/visible Haitian spaces

In spite of the historical and cultural ties existing between Haiti and France, the Haitian presence on French soil (including its overseas departments known as DFA) can hardly be defined in terms of the 'Haitian community'. This chapter looks for patterns of coloniality that inform the continuity of the historical burden. It explores how the French Republic deals with diverse communities forging livelihoods through family strategies, which are at the core of a transnational web based on complex patterns of inter-polarity, in particular from rural southern Haitian localities (La Colline, Fonds-des-Blancs) to French Guiana, Guadeloupe and Martinique. My multi-sited approach involves 'mobile positioning' (Massey 1995; Marcus 1998; Faist 2012) while talking to association leaders, elected officials and entrepreneurs to convey an understanding of an uneven geography of inclusion/exclusion in relation to memory and the nature of associative militancy in specific socio-spatial locations where creolization acts as a cement (Pointe-à-Pitre, Fort-de-France, Bordeaux, Strasbourg). I draw primarily on official sources on successive waves of Haitian migration in France and its overseas territories (Lebon 2001, cited by Mooney 2008; Joint and Mérion 2011; Insee 2014). My discussion leads to an effort to bypass a frequently hostile Eurocentric discursive construction of immigrants to uncover a rather dynamic and relational active citizenship of greater significance in the aftermath of the earthquake in Haiti.

Unlike the United Kingdom, with its strategy of decentralization and separation of its territories, for the most part France adopted an approach to decolonization involving assimilation through *départementalisation*. Beyond the legal framework of the French overseas departments (DFA), it is worth noting the recent debate about the departmental crisis after the forty-four days strike in Guadeloupe, then in Martinique in 2009. More than 161 years after the French Republic banned slavery in 1848 (Monza 2009: 179) and more than seventy years after the creation of *départementalisation*, I will first contextualize the situation that brought about the decision to implement this institutional framework through the law of 1946. Then I will address the question of whether a new geopolitical shift could be on the horizon and how the Haitian presence fits into such landscape.

The populations of four former colonies (Guadeloupe, Martinique, French Guiana and La Réunion) became integrated within metropolitan France and have been regarded as European territories since 1957. *Départementalisation* meant the double assimilation of institutions and people. After the Second World War, over seventy years ago now, these territories experienced a harsh socio-economic situation. They were marked by a high rate of poverty and of malnutrition, and an increase in the incidence of diseases. Housing did not have running water or electricity. Infant mortality was on the rise. It was therefore argued that 'assimilation was necessary to end social misery' (Belenus 2016). Yet, if the law meant guaranteeing civil, legal, economic and social equality between overseas territories and mainland France, it showed the need for a true assessment in line with the regional evolution.

This colonial doctrine, which originated from the concept of a unified French state, tended to deny public expression of any identity other than its own. In a broader sense, assimilation marginalizes all other identities for the benefit of an abstract citizen allegiance. After all, when the law was debated, it quickly became a long battle between those who supported measure they perceived to be a means of overcoming social hardship (Césaire, Bissol, Monnerville and Vergès) and those who feared the psychological and intellectual impact of a policy of assimilation (René Bénélus). Recently, during a debate organized by the Overseas Ministry, a consensus on very limited achievements was reached. The quest for republican equality with a strengthening of political autonomy and one's own cultural rights is difficult to reconcile with a coherent political framework. The expectation of a high standard of living has penalized the productive sectors, in particular by increasing production costs. Furthermore, the French West Indies have become isolated from their regional economic environment (Daniel 2016).

This is precisely where we can see the contrast with the multicultural policy implemented in the United Kingdom and how it plays an intricate role in creating (or failing to create) the right conditions for the formation of communities.

Memory and sense of belonging through spatial dynamics

During the wave of Haitian migration to OECD countries (1960s to 1970s), France was only a marginal destination for migrants, who mainly went to the United States of America and to Canada. Yet there has been a Haitian presence in France for a long time, the most significant one of all countries in Europe. Statistics published in the French census only give very little information on Haitians in Paris. Compared to numerous studies (doctoral and master's theses, official reports) about the Maghreb presence, very few academic works discuss the Haitian presence in France (Bastide 1974; Saint Louis 1988; Gazan 1990; Guillon 1994; Glaude 2000; Mooney 2008; Bechacq 2010). From these studies, some estimates can be made. In fact, Haitian mobility towards France dates back more than a century, long before the first big wave in the 1970s. In autumn 1959, a first group of about 144 students were awarded scholarships to attend Parisian universities by the French government. Among them were several people selected by the government of François Duvalier to become civil servants in the country (Kerns 2001: 63). Some of them went to the east of France to the city of Strasbourg, which has a tradition in existence of more than half a century of welcoming generations of Haitians, especially from the city of Jacmel (in the south of Haiti). This is the case of Syto, who settled in Strasbourg after completing his studies to become a surgeon:

> I was discouraged by my uncle at the Haitian consulate from coming to Chicago. He told me that I would lose my dignity while gaining some knowledge in a city so full of racism. So, my choice was to go to France and in particular to Strasbourg where my parents before him had lived. (Interview with Syto, physician, Strasbourg, 16 July 2010)

There was a significant wave of migration tied to study, in particular of medicine and economics. After finishing his first years of medical study in Port-au-Prince, Gary completed his training in Montpellier in 1953, financed by his well-to-do family:

> It was the good times! We were privileged to pursue our studies in this wonderful city in the 1950–1960s. About twenty of us shared memories right after World War II, at a time when there was a sense of euphoria about making a fresh departure on a new basis. (Interview with Gary, physician and member of the association Une école pour Haïti, Paris, 11 April 2008)

The immigration of this period consisted primarily of privileged individuals arriving in mainland France, but the closing of borders in 1974 came to upset the presence of the foreign population. Immigration of 'labour slowly emerged as well as family immigration. Further, the rate of foreigners who were salaried fell from 11.7 per cent in 1973 to 8.4 per cent in 1985, and to 7.3 per cent in 1988' (Frybès 1991: 83). It is difficult to understand the extent to which the migratory wave was caused by economic or political problems, because throughout the history of the country oppression has been a widespread and constant factor of exile. This is why some believe that the peak in the rate of emigration in the 1970s coincided with the emergence of an environment characterized by a lack of respect for basic socio-economic and political rights. For a nuanced understanding, migration needs to be appreciated through the prism of a Haitian society structured by emigration in all social classes. Society was under demographic pressure and its urban population continued to grow at a rapid pace while being exposed to an increasing brain drain. By the 1980s, several factors were at the core of a massive crisis of people arriving in boats in Miami. The political decision to eliminate black pigs as a result of some USAID reports of pig contamination or rumours of swine fever brought with it serious impoverishment for a large peasant class. Having lost their livelihoods, immigration at any cost became another strategy of family survival.

Migration dynamics is based on a whole system of extended families. This is the core of a larger strategy to facilitate ongoing access to different spaces in the Caribbean region and beyond. It allows progressive integration into any host society regardless of the institutional framework. As resources accumulate, socio-economic capital is also essential for the understanding of a family survival approach. It is often observed that instead of opting for a definitive return home or full assimilation in the host society, some migrants decide to diversify their choice of settlement. The student population in my corpus arrived between 1950 and 2000, in contrast to a smaller category of individuals whose arrival was based mainly on political reasons (1960 to 1991). None of my interviewees were born in mainland France. My field research about how living conditions are discussed in France and its overseas territories led to interesting observations regarding spatiality. Unlike in South Florida, there is no 'Little Haiti' in mainland France. Haitian communities are fragmented configurations supported by widespread relational networks. These are complex to discern, because they are multipolar and transnational in their practices. Family ties overlap with commercial connections that feed into a vast informal economy in the rural and urban areas of Haiti. There is marked polarization throughout Paris and its *banlieues*:

> Haitians are dispersed in little circles of friends and do not represent a significant ethnic group. A good example is what happened in the mid-1980s when a real estate opportunity emerged for some at Cergy-Pontoise (Val d'Oise). This explains the emergence of an important concentration of about twenty-four families originally from Gressier, Léogâne and Aquin. They own protestant churches and a few shops. (Interview with Siméus, financial analyst, Cergy-Pontoise, 4 May 2008)

Elsewhere in the Parisian *banlieue* of Seine-Saint-Denis, which is known for its significant immigrant population, the Haitian presence is visible. What is original about this geography is the change of the socio-professional profile of Haitians who have settled in the Hexagon in the past two decades. Pedro's journey symbolizes a timid integration into French society (in the years 1960–1970), and contrasts with Haitian settlement in overseas territories such as French Guiana, where this presence gained a growing visibility. There is a way in which accumulation proceeds through the iterative reproduction of coloniality: that is, through the reworking of hierarchies of socio-economic differences and forms of labour (Werner 2015: 13). This pattern is reflected in different Haitian communities spread in the banlieue or in the overseas. Depending on socio-economic circumstances, their mobility varies, as well as their choice of whether to remain outside mainstream society.

> Some of our compatriots often deny the existence of a Haitian community in France even when our families unite and form a community. Lots of energy and effort goes into the community being able to link these dispersed activities throughout different *banlieues* that will benefit the majority. Our directory, the first of its type launched in July of 1998, is certainly the expression of this desire to show others in the Haitian diaspora in America that we can also have our community. (Interview with Georges, entrepreneur and associative activist, Château Rouge, 20 February 2008)

Box 4.1 Between memory and silenced history

Siméus, twenty-eight years old, originally from Léogâne, is a financial analyst, and a member of the association Haïti Mémoire et Culture. This association was created in 2000, with the main goal of preserving the memory and cultural heritage of Haiti. It aims: (a) to teach the younger generation of Haitians abroad about Haitian history and literature; (b) to sponsor activities to promote Haitian culture within French society. 'What motivates me is a strange sense of urgency to make the younger generation understand the pertinence of preserving a Haitian history that is shared with France. Unfortunately, this history is not taught as it should be. This is why I call for an honest and positive dialogue between the youth born in Haiti and those born here in France to keep that memory going. This may help both sides heal and integrate better in the host environment'. Siméus enjoys discussing memory and history while insisting on the contribution of Toussaint Louverture to humanity. He sometimes goes against the grain by explaining events in a polemical way. For example, 'the Société des Amis des Noirs was not totally for the abolition of slavery, but just wanted to stop the slave trade'. Twice a month he hosts a three-hour programme on Radio RGB Cergy. This is an opportunity for him to play Haitian folk songs and to revive memories, so that the sense of belonging among Haitians in France and in the homeland remains intact.

Source: Interview with Siméus, financial analyst, Cergy-Pontoise, 4 May 2008.

In different cases, the profile of migrant workers tends to show a more varied level of competence and of formation without nevertheless overlooking the difficulties that are sometimes felt within the workplace.

> I do not see myself living here forever. I grew up; I have done my studies here with the feeling of never being accepted but just tolerated. Haiti needs me more than France. The problem has come from the financial crisis, which has cast doubt on my work prospects, as well as from the level of discrimination. I have experienced discrimination with my clients as a financial advisor and also in my relationships with certain colleagues. (Interview with Moise, retired architect, Paris, 8 May 2008)

Among my interviewees, six out of a total of thirteen were Haitians living in lower socio-economic neighbourhoods in Martinique. This presence appeared more female than elsewhere and is quite concerned with social difficulties, such as increased unemployment (30 per cent to 37 per cent in Zac Chateauboeuf-Bois Boyer) or precarious work in general, as in the case of Frantz, a native of Léogâne living in Terres-Sainville:

> I grew up in a sugar cane plant that produced rum. I remember lots of farmers working in this industry before Jean-Claude Duvalier came to power. Things got complicated later with an economic downturn and people started to leave the whole region. This is how I came to Martinique in 1978. The beginning was very difficult, but now I own my house and I manage my own agro-business enterprise. When I can, I recruit illegal workers in solidarity – 'moun Léôgane konn travay !' (People from Léogâne know hard work!) (Interview with Frantz, Terres-Saintville, 27 February 2010)

Across my fieldwork with a corpus of eighty individuals, my aim is to first treat the spatial inscription of the sense of belonging under examination. The meaning of bonds can be appreciated during the conversations I had with so many individuals whose perception of themselves and their host society resonates with whether or not they faced material obstacles. The level of engagement for a 'community' here and abroad (in the homeland) corresponds to a sense of belonging to a territory. The vast majority of my respondents recognize that they are different from both the mainland French and French West Indians. In that sense, they feel they belong to the Haitian nation with a vibrant past that remains present in their daily life abroad.

The ones who feel Haitian

In this category, we find mainly Haitians of the first generation who are profoundly nostalgic with vivid memories and traditions. Sixty individuals out of eighty feel Haitian and recognize that they are part of the nation even while living abroad. Those who define themselves as French are in a minority. There is a sense of belonging fuelled by a memory of a nation. This determines to a certain extent how they behave towards others in their host society. They position themselves according to who they want to be. Phil, originally from Aquin but living in Cité Cabassou, is a student and activist in an association:

> I arrived from La Colline (Aquin) in 1976 to join my dad who was living in Cayenne. I was thirteen years old. They made fun of me in school because of my accent but I adapted and gradually felt accepted. I have my family now, and I pass on my Haitian culture to them. I socialize primarily with my community. Today my children are adults and have more choices than I did: they can stay here or better they can go to the United States where I have cousins who are doing well. (Interview with Phil, Cabassou, 10 March 2010)

Interestingly, Bill, who came to Guadeloupe to pursue his medical studies in 2001, gives a different assessment of this territory. He describes an atmosphere that is a little hostile to its 'Caribbean brothers' while trying to understand such ambiguities:

> First, I am a Haitian citizen who is respectful of French laws. At the same time, I feel like a stranger in a society where welcoming foreigners is non-existent. I have felt stigmatized since my arrival which coincided with a polemical radio show by Ibo Simon who unleashed xenophobic speech against Haitians. This hostility was not directed at me personally, as I was working as a physician, but I still felt affected. I had never lived in a country that was not politically independent, and I began to feel doubts about my choice of coming here to specialize. I met so many neurotic people that I asked myself: is it the price to pay to enjoy a good quality of life compared to other Caribbean islands? My hometown is in the north-eastern part of Haiti and it is very poor. However, thank God, people have other problems that are not of a mental nature. Now, I follow the news closely, and I am looking at this mobilization against an expensive and unequal life. Finally, some are realizing that there is a profound societal malaise here. I discuss life in Haiti more with

> patients, and I sometimes feel more proximity. (Interview with Bill, physician, Pointe-à-Pitre, 26 January 2010)

Chris, originally from Jacmel, after medical studies in Haiti, then in Spain in the 1960s, arrived in Guadeloupe in 1986. He has a much more nuanced perception of the host society:

> I feel profoundly Haitian. I also feel that Haiti needs my help as a conscious doctor. As a union leader, I have many fraternal links with my Guadeloupean colleagues. We have got used to people after twenty years living here. It is a small island and we have Creole culture in common and a memory of great struggles. In that respect, I participated in the event of 'Lyannaj' by giving a presentation on Haitian history. This is a history made up of struggles to liberate the mind and the soul, much beyond the expectations of this ongoing strike led by the LKP movement (Lyannaj Kont Pwofitasyon), meaning an alliance against profiteering. Clearly, I share ties with those who sympathize with me and my country, Haiti. Of course, I believe that my sense of belonging is first with my compatriots, and then I have strong ties with Guadeloupean society. Everything I do is selfishly done in an indirect manner for Haiti. My wife often blames me for not being French enough. She does not realize how much of my potential has been used on French patients and almost thirty years of my life has gone by. I know that I am almost at the end of my career and I need to contemplate a different future now. (Interview, Chris, physician, Pointe-à-Pitre, 29 January 2010)

Adé, originally from Liancourt (department of Artibonite) has lived in the Parisian *banlieue* of Essonne since 1994. He feels in transit:

> I do not see myself retiring here. I have a few links here and there, but I give myself ten to fifteen years maximum. Then I'll look at other places to live elsewhere. Sometimes, I look at my life here and I feel that I will never be accepted. Perhaps Haiti and Liancourt need me much more than France does. (Interview with Adé, sociologist, Morangis, 3 May 2008)

The ones who feel French

Only four in my survey defined themselves as French. Are we in the presence of 'French originally from Haiti', in other words, assuming a pluri-civic belonging? The common characteristic of these respondents is that three of them were born abroad and the last one left Haiti when very young. Some events (natural catastrophes, hurricanes) have brought them closer to the rest of the 'community'. Lubin, whom I met at the Haitian Consulate in Cayenne, is a journalist. He came back overwhelmed from a short broadcasting assignment in Léogâne, days after the earthquake:

> I came in 1979 at the age of thirteen without ever asking myself questions about my family being originally from Miragoâne. I followed my passion for journalism, then, after the earthquake, I felt weird about my life. I devoted long hours in my profession going to any destination, except that this time I felt a sort of guilt about so much poverty in Haiti. I am grateful, because I see perhaps my luck to have lived in French Guyanese society. (Interview with Lubin, journalist, 8 March 2010)

The ones who feel Haitian and French

Although most people interviewed feel different from the rest of the population, twelve of them perceive themselves to be both Haitian and French. There is a desire to relate to the Haitian heritage and also to fully integrate in French society. For some respondents, there is no need to erase their cultural baggage, which is considered valuable and a benefit to navigate better in diverse spaces. Bijou is a member of a Haitian association and a tennis club in the twelfth arrondissement:

> I came with a student visa and then I found a good job and I feel integrated. I have friends and family and I spend time with my association Collectif Haïti de France while being very active in sports. Gradually, I got the best that society could offer me. I have not lost my Haitian roots. I feel both Haitian and French, even if, when I go to Haiti, I don't like the discussion about dual citizenship. I think that more needs to be done to accept the diaspora in its political rights to run for office. (Interview with Bijou, educator, Paris, 26 April 2008)

Unlike the Jamaican case, for Haiti the question of dual citizenship has been quite specific since the creation of the Haitian state and its various constitutions (the Constitution of 29 March 1987 is the twenty-fourth). Dual citizenship has been an issue in every electoral campaign since 1986, and in May 2011 it was formulated as an amendment of the Haitian Constitution just before Michel Martelly came to power. Articles (12,

12.1, 12.2, 13, 14 and 15) that forbade dual citizenship were abolished. Today the article reads as follows: 'No Haitian can use foreign nationality in the territory of the Republic of Haiti'. Some Haitian politicians are thinking about how to submit a proposal to the next legislature that would eventually ensure the inclusion of the diaspora during the next presidential election of 2022.

Other belongings

In a final category, there are individuals such as Melanie, who is twenty-five years old and who was born of a mixed parentage. She is at the University Antilles-Guyane and simply feels 'Caribbean':

> My experience in French Guyanese society does not cause me any problems because my parents prepared me to take what is best for myself in their cultures. They taught me to respect difference and to be proud of the Guyanese identity. This is why I feel like I belong to a larger group of people rooted in the Caribbean. (Interview with Melanie, student, Mana, 10 March 2010)

Quite well-informed about the lifestyles and customs in Martinique, Didi, who was born in Haiti of a Puerto-Rican mother and a Haitian father, apparently favours a plural belonging:

> When I arrived here, I quickly got married to a Martinican who loves Haitian culture. I then became the mother of a young boy and I am busy with my professional responsibilities. It is not easy to socialize with people I observe from a distance: their social level does not correspond to mine. Some are so well integrated in Martinican society that it is hard to tell if they are originally from Haiti. I am quite astonished to see so many Martinicans listen to so much Haitian music, our 'compas', as I was not so used to listening to it. (Interview with Didi, entrepreneur, Fort-de-France, 25 February 2010)

Some individuals were born of mixed Caribbean marriages, such as Daniela, a citizen of Guyana but born of Haitian and Martinican parents. She is the manager of her own company and does not hesitate to clearly show her position vis-à-vis Guyanese society:

I feel Haitian, living in a French country. I do not feel part of a group, I want to socialize with everyone without any prejudice. I respect everyone and I do not feel racist. There is a need to be accepted in the business world. At the same time, it is smart to make good use of the relational community network but it is wise to go beyond it to embrace a French republican culture and identity. It is possible to be French and not just Guyanese or Haitian. Just because I live among Haitians does not mean that I feel completely like them. In fact, I truly feel French and I feel more attached to Guiana than mainland France. (Interview with Daniela, born in Cayenne, Cité Bonhomme, 8 March 2010)

Table 4.1 Sense of belonging to Haitian nation

Sense of Belonging	Haitian	Haitian/ French	French	Other/Any	Total respondents
	60	12	4	4	80

Conclusion

In this chapter, our analysis has shown variations in the level of the sense of belonging within the population. As Table 4.1 demonstrated, distance and proximity affect the perception of the host society. Different migratory experiences are articulated in relation to a complex socio-political and economic situation in Haiti. At the same time, the context of *départementalisation* in the French West Indies has had some impact on the way the Haitian presence is perceived and at times manipulated. The Haitian revolution is often held up as an example of a failure that the French Antilles should not follow. In contrast to other independence movements, the Haitian Revolution was the result of an insurrection of the enslaved. Perhaps Aimé Césaire's drama 'La tragédie du Roi Christophe' contributed in an intriguing way to conveying more shade than light about the true nature of Christophe's kingdom and its legacy in the vision a progressive leader ought to have for its people. After all, Césaire, in his long presence at the French National Assembly, has never claimed to favour an independent path. Yet, the quest for a stronger cultural identity is exacerbated by efforts from some parts of the territories to place social and economic injustices in an historical continuum. From 20 January to 4 March 2009, the itineraries taken by a large mobilization coincided with sites of painful memory of labour resistance in colonial times and oppression by the central power. 'A victimised consciousness in the most vulnerable Guadeloupeans, to convince them that their conditions can never be overcome' was denounced, at the peak of a long strike of forty-four days in Guadeloupe (Lurel, French National Assembly 2009). Truly denouncing the reality of 'pwofitasyon' (exploitation, abuse) reflects the radicalization of protests affecting not only Guadeloupeans or Martinicans, but also immigrants whose experiences of belonging with respect to their host society are diverse. The discussion that follows in Part II explores in more detail the complexity of the evolution of the Haitian presence from weak visibility to greater activism in specific contexts of settlement.

PART II

The making of a transnational citizenship: A new dialectic between memory and civic engagement

On all continents, memory is necessary to construct the future.
One must never build on forgetfulness and on silence.
— Joseph N'diaye, 'One day in Gorée ... Recounting slavery'

CHAPTER 5

Imagining associative spaces in the French-British setting

The oldest Franco-Haitian association (1968) in Ile-de-France is Haïti Développement (HAIDEV), its activities mainly focusing on giving social and legal assistance to new immigrants. Ben, now deceased, was the dean of Haitian association leaders. He came to France in 1958. Discussing the title, he gave to the association, he describes how it arose from his perception of 'development as a social emancipation and equality of rights for those living in the host society'.

> You cannot help your brothers in Haiti without first organizing yourselves right here. To this day we continue to be one of the rare Haitian associations to be recognized for our ongoing actions in mainland France as well as its overseas territories in Guadeloupe and in French Guiana. We concentrate our advocacy towards the *sans-papiers* and the socially marginalized who are infected with the AIDS virus. (Interview with Ben, retired at the time of this interview, Paris, 23 April 2008)

The period between 1986 and 1994 saw three main political events: the fall of the Duvalier regime in 1986; the adoption of the 1987 Constitution, and with it the emergence of hopes of democratization and decentralization; and finally, the popular electoral mobilization of the Lavalas Party in 1990. The events marked a new beginning for many associations in terms of their mission and the degree of engagement towards Haiti. Pédro describes the context:

> Some of us who were already engaged in the struggle against the dictatorship realized that this was a crucial moment: the image of Haiti was totally degraded and it was necessary to promote another image with some hope. It was complicated to change my behaviour and to propose new objectives after struggling so long as a political dissident. (Interview with Pédro, retired union member, 30 October 2010)

With Haiti's changing socio-political context, new forms of associative engagement emerged among migrants with different motivations and objectives. The former cultural attaché at the Haitian embassy in France, Mrs Lafontant, confirms the change in orientation of a policy aiming at rapprochement among Haitians in France in her decision to order a report on the state of these associations:

> Our initial idea was to organize the very first official meeting with associations. This meant in my capacity of cultural attaché at the Haitian embassy in France, co-ordinating a platform between different circles of Haitians of France. Some refused to meet such and such a group due to a divergence of political views or other social considerations. Finally, the consensus was reached in collaboration with the 'wise' (les sages), meaning influential pastors and senior residents of the community. This first meeting took place in 1998. Preparatory sessions took place at the Consulate and a study was ordered to better assess the associative landscape. At the time, the report revealed the existence of about sixty Haitian associations in Ile-de-France. This first initiative was full of official symbolism and this certainly goes a long way towards explaining its overall success. (Interview with Mrs Lafontant, retired public officer, 30 October 2010)

Marie, formerly a part-time assistant (1997–1998) in the cultural unit of the Consulate, actively participated in efforts to unite the associations:

> I spent a long time with the association Haïti Développement while observing the fragmentation of the associational space. This initiative to unite had already been discussed before without materializing. Leadership was unfortunately lacking. As secretary, I gave assistance to the cultural attaché to facilitate the full organization of all associations. My contact with people originally from the city of Gressier (Haiti) residing in the *banlieue* of Seine-St-Denis, and residents originally from La Colline (Aquin) living in Essonne, helped the mobilization of the main communities dispersed throughout Île-de-France, at a time when the profile of Haitian residents was rapidly changing. (Interview with Marie, Paris, 30 October 2010)

Moïse leads Anacaona, one of the only associations that gives priority to defending against the violation of Haitians' civic and political rights in Haiti and in France:

> It is important to rethink development in terms of human rights and democratic participation. Many of us who left Haiti to flee persecution have reduced our links

> with the administration. However, the political upheaval caused by the fall of the Duvalier dictatorship could only lead to a re-engagement of some associations to consider the development of their activities so as to be directly connected to concerns on the ground throughout the places. (Interview with Moïse, retired architect, Paris, 8 May 2008)

The length of settlement in the host society, combined with the socio-professional profile, allows us to appreciate the nature of initiatives launched by the associations. Some have stopped sending money to family members and friends. Nevertheless, they regularly participate in collective local projects at home. Their contact with locals in France also guides them in hoping for a different vision of their role as 'in-between' actors who have the potential to bring about significant democratization in the homeland. In a way, what is at work is also a form of citizenship that goes beyond the narrow framework of national citizenship. Indeed, there is 'imaginative and communicative travel [...] [leading] to a range of complex identities and interactions' (Benson 2011c: 80). The degree of engagement varies according to the level of obligations an individual imposes on him or herself through membership of an association. 'An activist usually shows loyalty and consistency to ideals of obeying the objectives of the association even in difficult times' (Dumont 2007: 176). Nicot's activism has a collective and patriotic motivation. He arrived in mainland France in 1975, after a short stay in Martinique where he obtained political asylum. He quickly decided to tackle the lack of nurseries for Haitian parents looking for work. His engagement is based on his constant concern about integration in mainland France. The aims of his association CHAVE, created in 2004, are twofold: to work towards better professional integration in France, and to empower people in Haiti by providing technical assistance.

Box 5.1 Reinserting Parisian space through association

'Immigration often disqualifies individuals professionally speaking. It is difficult to refuse a job when you first arrive somewhere in spite of good diplomas. I did a lot of menial jobs until I decided to launch my own business. I was never able to progress in my profession to reach a high managerial position.

This is why I realized that entrepreneurship was the only way to make some personal gains. I changed career and worked in the import/export field.' Nicot's frustration at being held back in spite of good qualifications is clear. He shared many stories about colleagues who had also faced the same hardship. It is not just about being the victim of a labour system that forces so many to lower their expectations compared to their level of competence. There is a further structural component of the job market that prevents certain types of immigrants from obtaining access to more attractive sectors (Jones 2008: 145). Nicot became involved in the very first professional Haitian association in 1977. 'The objective to create the Association of Haitian workers born in France (*Association des travailleurs haïtiens élevés en France*) (ATHAIF) aimed at better professional integration, gaining confidence and accepting oneself. There was a need to have a network of information to give French classes, coaching and above all a literacy programme French-Creole. This was a fundamental way to help my desperate compatriots. This also gave me an interesting way of getting to know my community better. Being originally from Gonaïves, I came across many who were instead from the southern part of Haiti, and in particular from Aquin. In 2002, an association of people from La Colline (sixth communal district) selected me to become their member of honour'. Nicot became more familiar with the French labour legislation. His compatriots did not know much about the system. Once he realized that there was a problem in the method of the allocation of places in nurseries, he decided to get involved. 'In 1980, I created the first association France Immigration et Solidarité (FIS), with the aim of creating a parental nursery. I created the project that got the approval of the Ministry of Social Affairs. It benefited many parents in Cergy-St-Christophe where a good number of Haitian families were homeowners'.

Source: interview with Nicot, activist, Cergy, 24 April 2008.

I first became active as a member of the Association des chauffeurs de taxi (ATHP), one of the oldest professional associations that also organized socio-cultural activities. Then my objective in creating my own association CHAVE was to

> intervene to support the victims of my hometown affected by severe flooding in Gonaïves. I talked to some family members about the best options to support an inclusive structure in two spaces: here in Cergy where I could mobilize resources, and the other in Gonaïves where projects would emerge. It took a lot of discussion to agree on a long-term vision. A plan was conceived in two communes to install storage rooms for farming equipment (tractors, picks, spades) in order for people to work without expecting food aid. The initial idea was good and the association had a chance to boost people's confidence and to create a profound synergy between actions abroad and at home. (Interview with Nicot, activist, Cergy, 24 April 2008)

In terms of areas of intervention and domain of activities, my fieldwork is based on a detailed geographic listing of regions where associations are active in mainland France (Tables 5.1, 5.2 and 5.3).

Table 5.1 Type of Haitian associations in Hexagon

Type	Haitian	Total associations interviewed
1	Ties with MHAVE	3
2	Activism for the defense of citizen rights	5
3	Professional	3
4	Locality of origin – local development	17
5	Religious	2
6	Socio-cultural	4
	TOTAL	34

Table 5.2 Type of Haitian associations in the DFA

Type	Haitian	Total associations interviewed
1	Ties with MHAVE	1
2	Advocacy for human rights	8
3	Professional	3
4	Local development	6
5	Religious	3
6	Socio-cultural	6
	TOTAL	27

Table 5.3 Type of Jamaican associations in the UK

Type	Jamaican	Associations studied
1	Close to Jamaica Diaspora UK (JDUK)	4
2	Mediation – advocacy	6
3	Professional	3
4	Local development projects	5
5	Religious	2
6	Socio-cultural	2
	TOTAL	22

The results show a strong concentration of associative initiatives designed to support local development in several departments in the south of Haiti. All the departments are represented except the northeast, as I did not find any associations working in this particular region near the border with the Dominican Republic. Overall, associations conduct activities in the socio-cultural domain, and, over the past twenty years they have become more involved in local development projects. Associative members such as Eric, a member of Maison d'Haïti, based in the twentieth Parisian arrondissement, advocate to mobilize attention for Haiti and to do fundraising.

> I always thought that among us there was an opportunity to conduct some lobbying on behalf of Haiti to improve its image. Our cultural activities (Flag Day, Independence Day, Battle of Vertières, Christmas dinner) are meant to maintain the memory of a nation that is alive and well beyond borders. (Interview with Eric, medical doctor retired, Paris, 10 May 2008)

Among the associations which formed part of my fieldwork in mainland France and its overseas territories (DFA), I met an association whose orientation specifically deals with local development projects. It is dedicated to solidarity initiatives with the area of Liancourt (Haitian city of Saint-Marc). If the majority of associations I encountered are involved with socio-cultural activities, some are engaged more and more with socio-economic activities that can directly benefit sectors such as education, agriculture and healthcare.

Box 5.2 Spatial connection through peasants' solidarity in Liancourt

Adé arrived in France as a political refugee in 1993. He quickly made good use of this status by pursuing university studies, continuing all the way to doctorate level. Alongside this, he created the association ASOLID/Liancourt with very clear priorities. 'I am a native of Liancourt in the department of Artibonite known for its rich rice production. I used to teach the agricultural workers of Liancourt about the need for modernized farming practices that respect the environment. It was about my authentic identity tied to the land, and it is unfortunate to observe some young people who leave without thinking about coming back'. Adé continues his associative work simultaneously with the one based in his Liancourt village (Saint-Marc). He lives in the *banlieues* of Essonne, and the main goals of his association relate to local development in his area: 'My parents are landowners and my house was attacked due to a land conflict worsened by the political situation. Liancourt is a great domain of fertile lands in the Artibonite. Our family domain was more than 200 hectares and my association is autonomous. ASOLID regularly sends farming tools and provides produce for a local supermarket. I am supporting the construction of a vocational technical school called INATEC (National Institute of Technology) that provides classes on computing, electricity, plumbing and mechanics. We have a total of fifty-two students for now, but the campus is not yet complete. In the future, we envision cooperating with a municipality here in France or in the Caribbean to better ensure the long-term modernization of agriculture in the whole region'.

Source: Interview with Adé, sociologist of education, 3 May 2008.

Education in Haiti remains a priority for some associations working on the development of their place of origin. One unique advantage migrant actors in such projects possess is the possibility of acquiring knowledge in the host society and sharing it with the homeland. Most members interviewed are highly educated, and a few only carry a technical certificate (BTS). Some are attempting new disciplines depending on the job market, and even becoming entrepreneurs. This is how associative activity overlaps with professional ambition, and also offers an opportunity to 'show off' skills among old contacts in the homeland. This is a whole dynamic between a sense of belonging to an area and new roles played by the migrant in the living space abroad. Tet Kole is the oldest association in Guadeloupe, created in 1991 at Pointe-à-Pitre. It advocates for the rights of the majority of Haitians. Mira, a farmer since 1975, presides over this association:

> Tet Kole is an umbrella association that brings together the community in a hostile environment. If Haitians do not cultivate the land, can you imagine the Guadeloupean people doing it? So, we need to have a clear and honest conversation. (Interview with Mira, Petit-Bourg, Guadeloupe, 30 January 2010)

Marcel, originally from Cavaillon (south of Haiti), is a teacher who studied in Mexico before settling in mainland France, and then in Guadeloupe. He describes his vision of the usefulness of an association:

> Haitians living in Guadeloupe are much less respected than any other place throughout the Republic of France, which prides itself on being the country of human rights. One of our priorities through the association is a combination of three elements: political awareness, organization and action. It takes reflection and experimentation at each stage so we can efficiently mobilize and gain better visibility. (Interview with Marcel, teacher, Basse-Terre, 26 January 2010)

When I met Colette, my aim was to get a better understanding of the integration of individuals of Haitian origin in Guadeloupe. This moment coincided with a strike of forty-four days. Originally from Port-au-Prince, Colette works very hard to gain respect within the Chamber of Commerce and Industry. She recognizes her luck compared to her compatriots in precarious situations due to legal problems or a lack of qualifications. She describes a situation of people's 'malaise and discontent'.

Box 5.3 Activism through professional visibility in Guadeloupe

'At the Chamber of Commerce, I work on dossiers that opened my eyes to a critical situation and all sort of difficulties and discriminations against Haitians. I feel I have to work double to prove my potential'. Colette describes a Guadeloupean society that is very exposed to multiple identity crises besides economic hardships; she is close to the Fos Pou Konstwi Nasyon Gwadloup (FKNG) party, which claims independence.

> Her associative engagement is a political expression of 'what seems to be at the core of the problems in Guadeloupe: dependence. The adoption of modalities of social harmonization from metropolitan France contributes to a destructive impact on dynamic sectors of the local economy and a regular rise in salary costs. People coming from metropolitan France are seen negatively as profiting from an unequal and unfair system. They get better salaries than locals for the same type of work. The majority here is discriminated against. Haitians have to struggle even harder, and it is easy to lose the possibility to access serious posts in management.'
>
> Source: Interview with Colette, civil servant, 25 January 2010.

A socio-economic arrangement involving extensive dependence on external financing from mainland France puts significant pressure on the livelihoods and lifestyles of the population. At the same time, a portion of this population nevertheless considers a true form of autonomy to be 'unrealistic'.

> The island is kept in a state of dependence through financial transfers from metropolitan France under the principle of Republican equality. This only reinforces dependency even when there is a claim for more local autonomy through decentralization. How can Guadeloupe collectively aspire to more dignity as a proud people? In the Haitian case, the conquest for political sovereignty came first. Here the assimilationist claim favoured above all social and material modernization by projecting an illusory abundance that impedes all efforts of real autonomy. The right to self-determination is not included in the road map of the three senators who currently represent Guadeloupe in the National Assembly. (Interview with Colette, civil servant, Pointe-à-Pitre, 25 January 2010)

In the shadow of socio-economic demands and a powerful current of national identity politics, Colette describes hostile behaviour in particular towards Haitians, and even Haiti:

> Xenophobia is widespread after a year of difficult strikes led by the Guadeloupean LKP movement that mobilized public opinion. The reality is marked by great of inequalities. The relationship between Haitians and Haiti is very complex beyond remittances and frequent returns. There is pride in an authentic memory of an empowering revolution that can never be erased by a radio commentator like Ibo Simon (a radio host who frequently voices hostility towards the Haitian presence). In terms of co-operation, the ties exist but are still dependent on a few elected officials who

> show sympathy to the Haitian cause, as well as some intellectuals who are fighting in associations such as CORECA or in the union environment. (Interview with Colette, civil servant, Pointe-à-Pitre, 25 January 2010)

Colette's engagement reflects both a certain level of political awareness in Guadeloupean society and the will to find her right place. She questions the ambivalent nature of interactions with the Haitian community and she has no particular desire to return home.

> Since the earthquake, what type of solidarity do I see among Haitians themselves and with Guadeloupeans? I witnessed a solidarity that had never been seen before. Some mission repatriated victims at the CHU hospital (Centre hospitalier universitaire). On Tuesday, 26 January, we visited patients who had already been identified in a list from the Haitian consulate. It was not easy to co-ordinate efforts due to all sorts of suspicions and a lack of communication between the representatives and the community. The association Hagua Caraïbes responded to the first emergency assistance with the help of two nurses who volunteered at the CHU of Basse-Terre. They are preparing themselves for an eventual trip to Haiti. We also worked closely with the Red Cross to help the victims of this tragedy. What is asked of us is to identify patients evacuated from Haiti and their needs. Some of them may be eligible for asylum and others are only seeking medical assistance to return home later. (Interview with Colette, Pointe-à-Pitre, 25 January 2010)

Compared to Guadeloupe, associations are much less visible in Martinique, despite the long Haitian presence there. The question of how to define engagement in contrast to other territories was central in the interviews conducted in Martinique. Sony, originally from Léogâne, arrived in Martinique at the age of nine as a result of family reunification. He is the leader of the main Haitian association here, the Association des Haïtiens de la Martinique (ADHM). We discussed the overall settlement and the interaction with the people of Martinique:

> There are few associations compared to French Guiana and even Guadeloupe. This is perhaps due to the fact that some do not necessarily feel a common cause to fight for. The best place to meet lots of Haitians is during concerts or music festivals where many come to socialize with our Haitian musicians of 'Kompa'. Some musicians are well known here and keep going back and forth between Haiti and Martinique. (Interview with Sony, entrepreneur, quartier Ste Thérèse, Martinique, 26 February 2010)

Nevertheless, many of the individuals I met during my fieldwork do not enjoy as privileged a situation as Sony. Some like Monel live in populous neighbourhoods in Belle-Fontaine, struggle to make ends meet, and even operate in the informal economy. They do not necessarily enjoy relational networks:

> I am currently in a precarious situation in the construction sector. It has slowed down a bit and I am at the mercy of this fragile situation. My wife works in the hospital, and together we send less money than usual. We have not travelled home for the past three years. We want to participate in one of the associations, ADHM, which has the highest number of Haitian members. Usually when a band comes to town, this is when we meet more Haitians and we try to socialize with the community. (Interview with Monel, mason labourer, Belle-fontaine, 27 February 2010)

Regarding the organizational structure of the associations, it is worth noting the importance of the profiles of members and the association's mode of funding. The number of associative members is indicative of their growing role in relation to the homeland. Even when they are mainly composed of volunteers, their structures become better organized and go beyond a local dimension. How to explain this evolution of both the organizations and their functioning? Answering this question is fundamental to understanding associative dynamics and their interpolar character.

If the composition of associations reveals strong regional belonging and their own social practices, there are some nuances in their nature that are worth discussing. First, the associative space is not confined to a strict local nature but unites diverse elements of Haitian society. The example of the association Une école pour Haïti, created by someone from the commune of Fonds-des-Nègres (Sud d'Haïti), with the collaboration of a medical doctor from Port-au-Prince, shows how difficult it is to categorize by using the label of 'hometown', as is the case for the Haitian community in New York or Boston (Pierre-Louis 2001; Fagen 2009). This space plays the role of a site for apprenticeship for a peaceful co-existence that allows equal, horizontal relations of reciprocity and trust among individuals (Hamidi 2007: 358). While migrants' attitudes towards France are mutating, the change of direction is noticeable in the priorities of some associations. The reason for such a change concerns successive waves of new arrivals from different socio-geographic environments. The very first study done by the Agence haïtienne pour le développement local (AHDEL) in January 2000 underscored a strong tendency to participate in projects in the region of the south of Haiti. Gal, a master's student and AHDEL member, explains the significance of his research:

> The Haitian Agency for Local Development (AHDEL) wishes to identify diverse associative leaders in Île-de-France in order to better co-ordinate the first plate-form assembly of Franco-Haitian associations in Seine-Saint-Denis. Among the results, we noticed a growing interest in launching development projects in the places people are from. This is different from the period before 1986 when mobilization was almost solely focused on political struggle. (Interview with Gal, 13 April 2008)

Partnership is a viable route for associations only if they are able to go through a long and complex procedure of funding. They often become discouraged because of their limited knowledge of the bureaucracy they are dealing with. Among the thirty-four associations studied in mainland France, only six were able to design a funding plan to realize their projects. Among them, three are based in the city of Strasbourg. Syto, the main member of the Hibiscus association, found financial and technical support in the city. The Alsace region got involved very early in decentralized cooperation and twinning projects. The main structure in charge of this programme is well organized and is called IRCOD. The Regional Institute of Cooperation and Development (IRCOD) is an association created with the support of the Alsace region and is composed of 80 local municipalities; its purpose is to put its expertise at the service of local dynamics engaging with countries of the Global South. Syto told me that he approached IRCORD in 1987 along with the Urban Community of Strasbourg (CUS) to get long-term funding for his association and projects in his native city of Jacmel (south of Haiti). 'At the end, I managed to facilitate a twinning scheme between the city of Strasbourg and Jacmel. Irrigation projects and medical support are among the types of initiatives realized with my association Hibiscus' (interview with Syto, medical doctor, Strasbourg, 16 July 2010).

My interviews with Syto and with two members of other associations, Fidel from the Association Au Cœur des Bidonvilles (ACVB) and Amos from the Association Altitude 900, gave me more clues about the institutional context and associative projects. There is indeed some correlation between this context and the organizational structure of the association. Beyond the spatial capital that transforms 'the know-how to migrate' into 'know-how to circulate', the real challenge is to attain a level of associative engagement that mobilizes resources here to act there. Haitian associative action takes advantage of the institutional framework in which local authorities (i.e. of Strasbourg) are willing to provide the technical and financial resources. The next section will address the question of transferring social capital from the place of settlement to the homeland and how Jamaican associations emerged in the British setting.

Reconnecting Jamaican voluntary organizations and structures of mediation

In the context of multiculturalism, ethnic minorities benefit from an institutional framework that legitimizes their community belonging. This part analyses different processes of integration for Jamaicans within a transnational associative field. I examine associative dynamics by deconstructing the processes observed on the ground 'here and there' (Jamaica-UK). The articulation of memory coupled with a sense of belonging to a territory expresses itself through a detailed examination of several active associations. One such is Raffa in London and Birmingham, in which structures of mediation demonstrate how religiosity (Pentecostal membership) structures some community behaviour in relation to an Afro-Caribbean identity. These examples demonstrate the extent to which the transnational social network appears as an essential reservoir of local development.

The UK Jamaican community demonstrates a high degree of residential concentration. This has been of benefit to community organization initiatives ever since the numerous anti-discrimination struggles in the 1960s. A notable feature of the Jamaican expatriate communities observed was the proportion of strong individuals to associative practices within the population studied compared to the Haitian case. In other words, I refer to the number of people getting actively involved in associative practices. I explore whether the specificities of the migration experience within the British institutional context determine this engagement. Among the different types of associations identified, those working on civil rights and opposing discrimination were the most numerous.

One such is the Derby West Indian Association, which has been in existence for over thirty years in Derby in the East Midlands. Kate explains how this association continues to serve the Jamaican community, most of whom are Seventh Day Adventists originally from the parish of Manchester, Jamaica:

> This was a time (the 1960s) when hostile and even racist attitudes were particularly violent towards immigrant populations. The West Indian association provided us with a space to unite and claim our rights. We needed to organize to challenge discriminatory practices about housing. According to a long British tradition, any local association could organize itself on a community basis. In the end, for us from the Jamaican parish of Manchester, it was not complicated to get together around common values in an environment where it was difficult to evolve. (Interview with Kate, teacher, Derby, 10 February 2010)

Box 5.4 Pan-African heritage and collective memory

ACSHO (African Caribbean Self-Help Organization) is one of the oldest Jamaican associations, and its main focus is awareness and activism in the Afro-Caribbean community. It mainly serves as a mediator in protecting the civic rights of ethnic minorities. Kwas is the leader and is respected by the young people of the district. His association's initiatives are well known by the Birmingham City Council. In so doing, it has gained legitimacy to denounce frequent problems, such as police violence and abuse. Solidarity is the foundation on which associative networks rest. 'In a hostile environment, solidarity with all those of African descent throughout the diaspora is vital. It allows us to raise the level of awareness about a sad reality: they no longer need us. Many among us suffer from mental illness or are even locked up in psychiatric centres. My solution is to be conscious of yourself and to look for your space, if not it is useless to look for acceptance as Black British which is nonsense. My conviction is that the struggle needs to play out in the political arena, and my engagement is based upon my deep memory of the strength of the Afro-Caribbean people as a whole. A figure like George Padmore advocates a Pan-African revolutionary philosophy enlightened by a vision of self-determination for more Afro-Asian solidarity. Each May, it is African Liberation Day)'.

Source: Interview with Kwas, associative leader, Handsworth, January 2010.

Kwas' belief system is pan-African and his activities are geared towards a set of initiatives beyond Jamaica. Besides the civic struggle and problems of discrimination and racism, his association organizes regular educational sessions for young people. He launched a fundraising initiative to build a place designed to deliver services and professional training for the community. His statements reveal the socio-historical background to the wave of migration to which he belongs (1959), and the British policies towards this generation. He is from this older 'Windrush generation', whom Winston James (1993: 256) described as having inherited and endured a form of 'migratory traumatism in the wake of the boat SS *Empire Windrush*, which in 1948 brought the first Jamaicans to the UK. Such trauma seems to have been at the core of the initial associative mobilization, which was based on a common awareness of a painful colonial past and its imbrications with the continuing iniquities of a neo-colonial present. Kwas is an icon of this epoch, and his narrative shows the pertinence of the context for explaining migratory behaviour and associative engagement. Foner (1985: 712) underscores the weight of the context of residency that becomes pertinent in a comparative approach.

Unlike France, the British multicultural context appears to facilitate the mobilization of ethnicity as a political resource among the minority communities. Yet, the pluri-ethnic character of the Caribbean population points in a different direction. Indeed, a study done at the University of Warwick (Owen 2009: 63) reveals a high percentage of children born of mixed marriages, and, as a result, there is a positive trend for the younger generation to fully integrate into the British model. Could this mean a lower capacity among the Caribbean associations to mobilize this ethnic dimension in contrast to those of Indo-Pakistani origin (Lacroix 2012: 75)? The structuring of the Jamaican associational space appears to be characterized by three elements: the possibility to acquire professional experience via associative engagement, religious belonging, and the strong sense of belonging to a Jamaican nation (a 'sense of Jamaican-ness').

Box 5.5 Traumatic memory of childhood in Jamaica

Jocelyn, a sociologist, decided to leave her career in teaching in order to work full-time running a community centre led by the Caring Opportunities Partnership Equality (COPE) association in the disadvantaged neighbourhood of Handsworth. 'COPE was created in 1995 to support a programme for women victims of premature pregnancy and for young people affected by failing school. This structure has rapidly expanded to welcome young people who are victims of mental illness. I was touched by so many of them coming out of Jamaican dysfunctional homes. This is when I decided to stop my teaching at the university to embrace an activity that is more valuable in my eyes with my Pan-African ideals: to help oneself to become autonomous'. Jocelyn decided to leave the university space that she considers too distant from and inadequate for the needs of her 'community'. The association has become for her a space marked by frequent interactions and exchanges with members of her community. She went through a difficult adolescence, during which she was emotionally very distant from her parents. So, her associative engagement fits an emotional need: 'life is difficult for these young people who too quickly abandon the school system that does not offer them learning that is adapted to their tough environment. They need to be mentally re-trained to better lead their life on their own. Medical therapy is not sufficient on its own. Social aspects are at the core of the problem: isolation, exclusion, poverty. My approach consists in sharing my experience to inspire them and also empower them'.

Source: Interview with Jocelyn, sociologist, Handsworth, 22 November 2010.

The way professional competence is used to enhance associative engagement demonstrates marked specificities in the British context. Indeed, it is in line with the multicultural policy that encourages minorities' accountability via the support of associations. Mussa, a Rastafarian born in London, first trained as a plumber in the district of Soho, Birmingham, leads an association that mediates between young people and the local authorities.

> Our aim is to become autonomous by making proper usage of financial resources. The Rastafarian culture is stigmatized and this often creates difficult interactions between Caribbeans or with the authorities. So, we intervene in giving training to our young people and to guide them in taking care of themselves. (Interview with Mussa, Rastafarian artist, Soho, 6 November 2010)

I observed a great deal of formality in the statutes, rules and internal practices of the Jamaican associations. In different cases, the profile of migrant workers tends to show a greater variety of levels of skills and training, with difficulties sometimes felt within the workspace (Table 5.4).

Table 5.4 Overview of Jamaican community engagement

Type – Organization	Year Established	Activities	Leader's Stay	Leader's Education	Leader's Occupation
1. JDUK					
ARRR	1983-Birmingham/ Kingston	Support to Return Residents	34	College Diploma	Entrepreneur
Friends of Margaret Hospital	1992-Birmingham	Support medical Equipment	50	University level	Business Manager
2. Advocacy/ Mediation					
ACSHO	1964-Handsworth	Mediation/ Social activism	36	College Diploma	Activist
Derby West Indian Association	1981-Derby	Cultural events	51	University level	Academia
3. Professional					
NAJ	1978- London Birmingham	Education	49	University level	Nursing/ Consulting
NBP	1996-Birmingham	Minority BME Support	44	Management School	Business Entrepreneur

Type – Organization	Year Established	Activities	Leader's Stay	Leader's Education	Leader's Occupation
4. Local Development					
TORA	1998- Walsall	Disaster Relief Deeside	50	College degree	Retired Council Rep.
Erma Jamaica Fund	1998- Birmingham	Disaster Relief Portland	52	Nursing school	Social work/ Nursing
5. Religious					
RAFFA	2006- London	Knowledge Transfer	2nd generation	Business School	Practitioner/ Consultant

On paper, the organization may seem perfectly structured, without however being very active on the ground. Moreover, an associational structure such as this, marked by hierarchies, poses the risk of the multiplication of posts (president, secretaries, etc.) solely for personal ambitions. This can get in the way of the content, orientation and objectives of the association. Finally, my interviews and meetings also reflected the way a small clique can exhibit an excessive and disproportionate influence instead of having a broad collective and representative leadership of the will of a larger majority. These observations raise the question of community leadership, as reflected in the intervention of one of the secretaries of the Jamaican Diaspora Unit (JDU) during a preparatory meeting of the national diaspora conference:

> Since 2005, I have socialised with members of this assembly. I currently represent the secretariat of the West Midlands. As such, I talk on behalf of about forty members of this body. The constitution states that the national council of the Jamaican Diaspora Unit in the United Kingdom (JDUK) must be composed of constituents' members. Yet, those who have authority did not inform the Jamaican government and refuse to discuss issues of concern for the diaspora. This goes to show the structural incapacity of our organisation. Others who are present today will have to select a council representative without knowing the legal procedure. It is grave that the JDUK acts badly as an institution whose modes of functioning remain obscure. (Statements of the representative of the JDU, Birmingham City Centre, 30 April 2011)

Most Jamaicans questioned are engaged in transnational practices of a socio-economic nature, and in particular with remittances or even investments in business. Out of twenty-eight individuals, only two of them do not belong to any association. One of these two, Victor, avoids such membership to preserve his 'objectivity' as a teacher:

> There isn't any conflict in myself and it is a fact that I am Jamaican 100 per cent. I don't seek to discourage activists who mobilize around a different identity, but I can understand when people are curious to know why I do not get involved in an association. I believe that I have made a clear choice to be in an academic approach and not an activist one. I know that I do not wish to mix the two behaviours. I teach and I work on numerous publications. This is my modest contribution. When I write, I put all my Jamaican soul into my activity and my essays and books are certainly inspired by my culture of origin. Sometimes I decide to support certain organizations. You have different roles for different people. (Interview with Victor, entrepreneur, Aston, 28 January 2010)

Conclusion

The existence of a network affects the way ties between different poles of the migratory space work. With the rapid evolution of migrants' profiles, more specific questions on the nature of long-distance engagement need to be discussed.

How can social and spatial interaction linked to projects implemented by these delocalized actors best be appreciated? Could the sense of belonging migrants establish in the associative context contribute to the emergence of territories of a new type in their homeland as well as in the host society? If so, is it possible to consider a participative dynamic that is introduced directly and indirectly by them via a complex network beyond borders? To respond to these questions, I have thus far described the Haitian and Jamaican space through their main common characteristics, in particular in terms of activities carried out by the associations. This engagement can be inscribed in the continuity of an individual's journey. It can also be seen evidence of a will to adapt to French society (in the case of French Guiana), or it may reflect a broader perspective of attachment through memory and the need to return home (the Jamaican case). Shared interests contribute to feeding solidarity and a multiplicity of belongings. The next chapter will discuss in greater detail another angle that is specific to the two populations studied: the extent of a religious transnationalism in long-distance associative engagement.

CHAPTER 6

Community engagements, solidarity funds and politics of memory

Much of the literature that deals with diaspora politics makes clear reference to electoral voting system and citizens' rights (Laguerre 1984; Portes 1996; Lafleur 2005). Less often, the focus of the debate is on the ways in which collective memory is part of an instrumental political process for both sending states and expatriate communities abroad. This chapter discusses the evolving nature of the relationship between the Haitian and Jamaican nation states and their overseas communities. My aim is twofold. First, I seek a better understanding of an underappreciated link between migration and long-distance engagement in area studies. Second, I argue that the merging of the politics of memory and the activism of transnational communities has the potential to impart empathy in a broader geopolitical context. A primary finding explained in this chapter is that regarding a decisive difference between Jamaican and Haitian associative transnationalism: the political dimension of the migrant's associative actions relative to their relationship with different institutional interlocutors, whether state, local or international (Organisation de la Francophonie (OIF) and the Commonwealth).

Remembering Haiti long before the making of the Francophonie

> Great or small, strong or weak, near or far, white or coloured, let us never forget that one day we may be somebody's Abyssinia.
>
> — Alfred Nemours

At the plenary session of the League of Nations (SDN) in 1935, Alfred Nemours in his role as representative of the Haitian delegation condemned the aggression of Italy against Abyssinia (Ethiopia). Many years later, Senator Emile St Lot's action, during the vote of 17 May 1949, was decisive in the rejection of the Bevin-Sforza partition plan of Libya. These two diplomatic events remain vivid in the Haitian collective memory. They attest to the singular positioning of a country of limited material resources that is nevertheless; capable in the concert of nations of defending its ideal of liberty beyond its borders, and especially at times when great powers are behaving against this value of humanism (archives of family St Lot; Dorsainville, 2006: 32).

For many years, migrations have played a key role in international relations. In the ongoing debate regarding the relationship between migration and development, the treatment of migration from the perspective of the countries of origin and diasporic policies has been predominant in the literature. In the era of globalization, characterized by an open extra-territoriality of socio-political, economic and cultural relations, states are undertaking to approach their overseas populations through a politics of nationality (or dual citizenship), of representation or voting at the level of the planet (Dufoix 2010: 17). For how else might one define a diasporic politics by a 'diasporic state'? By this notion of a 'diasporic state', we mean, one that 'considers dispersion as a mean of ensuring the survival of the nation and even its economic development via the usage of a migratory capital' (Thiollet 2007: 290). In pursuing this enquiry, the discussion now changes direction. From the prior exploration of the political opportunities offered by the host country, the mode of organization of the association and the sense of belonging already demonstrated as common factors of Haitian and Jamaican associative transnationalism, the discussion now turns to the institutionalization of interlinkages between state and migrant associations based on their action in the political space.

Introduction

> We must begin to rethink our problems and their solutions in our own Haitian manner [...] In Europe, in France, the welfare state is everywhere. In Haiti, the total weakness of the state may be our chance.
>
> — Frère Francklin, 1997

The foreign policy of states today takes into greater account migrants as new actors on the international scene. They represent an evolving reality that calls for more research from the standpoint of the relationship between the host state and the sending state. The Francophonie as instantiated through the Organisation international de la Francophonie (OIF) is of interest in this section in terms of the way it interacts with three actors: migrants' associations (France and its overseas territories in the Americas, DFA), Haiti and France. I use 'Francophonie' for its political meaning as an institution and a geopolitical space in a globalized context, constituted since the first Paris summit of the OIF, in 1986 under President François Mitterrand. Since its inception, Francophonie has developed a specific relation with Haiti, the sole independent Caribbean country since 1804. Haiti official language became French but also Creole under the 1987 Constitution. The term 'francophonie' was coined by the geographer Onésime Réclus (1837–1916) in his book *France, Algérie et Colonies*, and it was only in 1962 that it was defined in the dictionary Larousse as 'a collectivity constituted by people speaking French'. Since its inception, the institution of La Francophonie emerges out of the decolonization process under General Charles de Gaulle in France and the efforts of Leopold Sedar Senghor, President of Senegal and of Habib Bourguiba, President of Tunisia. This geopolitical space has not ceased to extend its ambit beyond this, into countries that were not

French colonial possessions (Gazeau-Secret 2010: 39). At the twentieth conference of the heads of states and governments of member countries of the Organisation Internationale de la Francophonie (OIF) in Dakar (November 2014), the membership included eighty-four states and governments: fifty-four full-fledged members, four associated members (Cyprus, Ghana, Qatar and New Caledonia), and twenty-six observing countries. In total, the global francophone population numbers approximately 220 million. From OIF data (2009), the population covered by the OIF represents 3.2 per cent of the world's population.

Haiti was one of the founding members of the Organisation Internationale de la Francophonie (OIF). As one of its region's most economically constrained countries, it is of interest to explore the effect of OIF membership on Haiti. Historically, Haiti also the only nation in the world created as a result of a successful mass liberation from the system of modern Atlantic slavery. It is a country built on the ruins of the slave system. Haiti's history therefore embodies an enlightening human experience at a time when the United States of America were slave states. Without a profound understanding of the complex process of its national formation, it is difficult to appreciate the differences between Haiti and the processes of decolonization of other countries that are part of OIF.

Jacques Chirac was asked by a journalist, in reference to Haiti, 'What does France as a rich country do for its former colony? He replied:

> Haiti was not, strictly speaking, a French colony, but we had for a long-time friendly relation with Haiti to the extent that we share in particular the usage of the same language. And France has been having an important co-operation with Haiti that keeps growing even more, and it will continue. (10 March 2000, Jacques Chirac, Pointe-à-Pitre)

These statements are quite symptomatic of the collusion between memory and history, especially given that they come from a former French president. They highlight further highlights the complex relationship that a former colonial power entertains with a former colony. It is pertinent to remember that Haiti is the only nation in the world that was built out of the ruins of a slave system with the key ideal of liberty or death. Unlike what is often mislabelled in the public opinion and even within the academic sphere,

Haiti does not fit a classical process of 'decolonization' as it is the case of the majority of countries in francophone Africa for example. Instead Haiti offers a unique perspective in the field of international relations because of its active role in the creation of francophone institutions (Duhaime 2010: 126). In what follows, I will discuss a number of weaknesses and ambiguities about what Francophonie means for Haiti and the Haitians of France.

The francophone presence in Haiti is reflected through the extent of engaged initiatives launched by diverse structures in a decentralized framework. Given the challenges of the Haitian educational system in particular in the rural areas, I argue that the time has come for a critical assessment of Francophonie initiatives at municipalities. It is a largely créolophone milieu. I use data from my fieldwork in the department of Grand'Anse (2010–2011). It has twelve communes and seventy-nine communal districts. In this space, OIF has a total of three schools (called '*écoles fondamentales d'application*') with some pedagogic support centres (EFACAP). In the commune of Abricots, the association of mayors of Grand'Anse (AMAGA) receives institutional support from the International mayors' association of Francophonie (AIMF) which aims at facilitating the realization of development projects in the South. The francophone practice can serve as an engine to encourage Haitian local municipalities' engagement. The university agency of Francophonie (AUF), the regional conference of chancellors and university directors of the Caribbean (CORPUCA) are institutions representative of the Francophonie support to the Haitian higher educational system. Yet, as a cultural, consumerist and associative space in connection with the French language, La Francophonie falls short of expectations on the migrants' part.

One of the OIF structures, the Agency of Francophonie Universities (AUF), participated in the regional conference of university presidents, directors of academic institutions in the Caribbean region (CORPUCA), on 25 and 26 February 2010 at the University APEC of Santo Domingo in Dominican Republic. The raison d'être of this conference was to contribute to the development of higher education in Haiti. In this domain, the will for La Francophonie to get involved in the socio-educational landscape post-earthquake became quite clear in its actions on several fronts. President of Université Antilles-Guyane (UAG), Pascal Saffache, announced his solidarity with fifty-two Haitian students who survived the Institut Aimé Césaire's destruction by welcoming them temporarily to the Master management of territorial municipalities at the faculty of Law and Economy, at the Campus Schoelcher (Martinique).

> The University Antilles-Guyane (UAG) plays a diplomatic role and of co-operation in the Caribbean. This role of co-operation is growing because of the necessity to provide higher education that is affected by the earthquake in Haiti … in a long run we will welcome fifty more students on the Martinique campus. (France-Antilles, n° 12945, 2 March 2010)

Francophonie has long been the object of a debate where French language is perceived as a marker of particular social status in Haiti, besides the other official language, Creole. In the host country, linguistic practices continue to occupy an essential function in the constitution of collective identities and individual vibrant memory. A study carried out on the second-generation of children of Haitian immigrants confirms the complex integration process in the French Republican schools (Luzincourt 2007: 63). The usage of the French language is a dilemma for Haitian households in France: should Creole be maintained so that the youth do not lose their identity and preserve elements of their cultural heritage of the ancestral land? For some settled in France to express themselves in French means a factor of valorization, Creole is more reserved to the private family sphere. Yet, for a much younger generation born in Haiti or even in France, of Haitian parents, English appears to take over in a créolophone and Anglophone space that cuts across communities spread in the French Antilles to the Parisian banlieues. Nevertheless, for a younger generation born in Haiti or even in France, of Haitian parents, English seems to dominate in a créolophone and Anglophone environment among Haitian communities abroad in French Antilles and in the Parisian banlieues, apart from the proximity of some of them to the Afro-American pop culture. All of these influences contribute to a declassified reaction to French as a profound identity and linguistic vehicle thus far. Lunie's experience in France since the age of ten, illustrates the ambiguity of the usage of her maternal tongue (Creole) and of French acquired in the French schools.

> In 1987, we were just five families of Haitian origin settled at Chaville in Hauts-de-Seine. As the only child, my parents had a great deal of ambition for me in pursuing my university studies in law. My father, an opponent to the Duvalier regime in Haiti, was forced to cease his activities in the courtroom in Port-au-Prince and to flee in exile. I chose to study European international law at the University René Descartes, Paris V. During several stays in Haiti, my parents encouraged us to speak French with no accent so that around us we could distinguish ourselves to the rest of the Haitians. Yet, as soon as we were away from the capital city to visit rural towns, we became immersed in a créolophone ambiance marked by strong American influences. I am working on cultural projects that put our belonging to the créolophone and francophone world. At the municipal team at Chaville, as a member of the right political party, I appreciate the challenge for us to fight for cultural diversity in the political agenda. (Interview with Lunie, Chaville, 1 July 2008)

Her political engagement on different issues (defence for the 'sans-papiers', affirmation of a culture francophone and créolophone) reflects the 'development of an active citizenship as direct participation to public affairs' (Neveu 2009: 145). Lunie has the 'know-how to navigate' in multiple networks and she is an activist for cultural diversity that is inclusive of the créolophone dimension shared with the French Antilles neighbour of Martinique, Guadeloupe and French Guiana. This is precisely the type of cultural capital in need of better investment by the Francophonie towards the youth in particular. This example shows the utility of migrants' associations as bridges or mediated forces in the cultural francophone space. The logics of transnationalism as unique territory of social action where migrants evolve in different cultural and social systems take a pertinent signification in the frame of Francophonie. As a part-time talk-show radio host of Espace FM, Lunie uses her appearance once a month to spread her message for a proactive role of the youth. She believes that cultural diversity can best unite the créolophone community whether from the French Antilles or Haiti on a platform of common neo-colonial memory and identity:

> The possibility for young immigrants to participate in a common cultural and linguistic universe remains unexplored and it is unfortunate. There is a frequent tendency to see Francophonie as a space reserved to well-known writers and poets. Yet, a majority of the youth is actively looking for better access to the artistic domain in the French Antilles and could gladly be part of art workshops. At Chaville, there is a need for painting sessions, slam training and our music Kompa to be better known

> in the French society. It is possible to raise funds within the municipality with private entrepreneurs interested in helping the youth positive engagement. Summer workshops is one idea to rally musical bands and young artists exchanges in the least prosperous parts of the Hexagon and of the French Antilles. (Interview with Lunie, political activist and artist, Chaville, 1 July 2008)

This conversation with Lunie reveals how migrant associations can serve as efficient tribune in the francophone cultural space. Literary events or poetic evenings organized by Franco-Haitian associations such as the 'association *Pour Haïti* in Paris, facilitate interaction between local cultures where writer-migrants are from and culture of the place of settlement. The logic of transnationalism as a unique territory of socio-cultural action where the immigrant population move in different cultures and social system has great meaning in La Francophonie space. Moreover, back-and-forth movements are opportunities for these associations to nurture exchanges of material, immaterial and ideal goods with La Francophonie. The participation and financial support (approximate 5,000 euros) of OIF represented by Mme Moreno, to the associations of the Plateforme d'Associations Franco-Haïtiennes (PAFHA) on 30 October 2010, is the first time in the history of this associative plat-form in the Parisian banlieue of Seine-Saint-Denis. In the years ahead, it will be interesting to track on a long run the real significance of such an OIF initiative, given that this participation came out of the well-orchestrated visibility of La Francophonie in the post-earthquake context.

Jamaica and the Commonwealth: Constitutionality, minority rights and post-Brexit

Contrasting with the Francophonie, the Commonwealth is discussed on two major aspects: its relationship with Jamaica and the work done by the Ramphal Commission about the theme of migration and development in a pre- and post-Brexit world. The Commonwealth, a multicultural and multi-confessional geopolitical space as a result of the British Empire decolonization, is a body of fifty-three states. At the Imperial Conference of 1926, the Balfour Report underscored autonomy of different member states, equal in status, united by the same loyalty to the British Crown. What remains to this day is to a certain extent, a form of allegiance to the British Crown more than a common language as it is the case within the Francophonie.

At the third Conference of the Jamaican Diaspora in London (June 2012), the fiftieth independence anniversary of Jamaica coincided with the jubilee year of the Queen of England, a heritage of the British colonial era. The then Jamaican Ambassador gave a vibrant speech alluding to the possible revision of the Jamaican constitution in order to become a Republic. Considering that the constitution is not 'inherently' Jamaican' nor representative of the expectations of the Jamaican people, she placed her party – the People's National Party (PNP) – as a nationalist front that defends the republican option for her country without calling into question its belonging to the Commonwealth as a broad geopolitical institution.

> The founders of the Jamaican Constitution in 1962 chose to keep the Queen as the Head of the Jamaican state. At the time, the dominant feeling was that the cultural and historical association with the United Kingdom justified the preservation of this tie at the expense of the republican option. This feeling in favour of the monarchy system in 1962 was so strong that the founders of the Constitution decided that the Queen would remain an integral part of the constitutional framework. Hence the Queen of England is the Queen of Jamaica, and it is mentioned in the Constitution: 'Elisabeth II, by the grace of God, Queen of Jamaica and its dependencies and territories, chief of the Commonwealth.' (Mrs Aloun Ndombet Assamba, High commission of Jamaica in the United Kingdom, 16 June 2012)

In 1970, Guyana (independent since 1966), or in 1976, Trinidad and Tobago (independent since 1962) both decided to revise their constitutions (no continued recognition of the Queen of England as Head of State), and they both became republics. Besides the Jamaican legislation that recognizes the nationality of its citizens who can reclaim their filiations, academic works have been done on the procedure to engage constitutional reforms to obtain a republic status (Munroe 1983). The Ambassador's speech made references to the Prime Minister, Portia Simpson's statement in January 2012 about the decision to go further on the road of institutional independence through a project of 'Republic'. On the contrary, the Jamaican case constitutes a challenge in regards to the Jamaican or even British-Jamaican public opinion and associations often acting as lobby groups among British authorities. The sense of belonging of a vibrant memory varies greatly depending migrants' experiences throughout the United Kingdom and at large within the Commonwealth. The structure Jamaican Diaspora United Kingdom (JDUK) serves as a framework to relay a discourse of a political nature on the relationship between the Jamaican state and the United Kingdom, but also between the British state and its citizens of Jamaican origin. The will of the Jamaican state to make its diaspora an actor of its foreign politics can be put in parallel with the attitude of the Commonwealth which recognizes its debt towards migration and celebrates the cultural, religious and institutional diversity of its members as a major point of its overall politics.

Jamaican Ambassador, Ndombet Assemba, discussed throughout this speech the dilemma posed by the current Jamaican constitution, one of the legacies of the British colonial era. Given that the constitution is not intrinsically Jamaican or representative of the expectations of the Jamaican people, she put forth the nationalist line of her party, the People's National Party (PNP), at a symbolic moment of the fiftieth independence anniversary of Jamaica. She defended the Republican option for her country without calling into question belonging to the Commonwealth as a geopolitical space. What might be the impact of this political discourse conveyed by a Jamaican Ambassador in front of a mixed Jamaican and British Jamaican audience and even some diplomatic representatives? This interrogation opens a broader debate far from being exhausted for which direction Jamaican associations should take. Some associations clearly see their role as pressure groups on British authorities especially on immigration issues. The same line of conduct can be documented in another settlement space such as Canada or the United States. The year 2012 marked the Jubilee of the Queen of England, as well as the fiftieth anniversary of Jamaican independence. That said, the sense of belonging to the Commonwealth varies depending on the migration journey of Jamaicans who participate in British political life. I asked the question of the influence of the sense of belonging to the Jamaican nation

(Jamaicanness) in order to better appreciate the extent of long-distance associative engagement and what it means in terms of local development projects toward Jamaican parishes.

Through the structure JDUK, the fiftieth anniversary of Jamaican independence served as a conduit to spread an essentially political discourse on the nature of the relationship between the Jamaican state and the United Kingdom and, also, to a broader extent, between the British state and British citizens of Jamaican origin. The political will on the part of the Jamaican state to conscript its overseas communities into its foreign policy can be articulated in terms of a broader collaboration with the Commonwealth. This is an important means of demonstrating a sort of coherence in accepting the positive aspect of the great diversity of this geopolitical space. Yet still not much is written about how the post-Brexit Commonwealth policy will evolve, which often involves deliberately forgetting about the painful past and driving memory at the heart of a more nuanced understanding of the activism of transnational communities ready for more participatory, bottom-up partnerships with UK institutions.

One of the components of the approach taken by the Commonwealth is discussed under the prism of the articulation between migration and development using some of the work produced by the Ramphal Commission. This institution created in 2009 under the leadership of Jamaican Prime Minister P. J. Patterson, produced a key report of three axes (Thomas-Hope 2011): lucid management of migration as a powerful human force beneficial to development and governance; impact of new technologies of communication on migration; cooperation between states for a smart and just transfer of human resources to prevent brain drain among member states. The commission is however, dependent on the cooperation and financial support of the Commonwealth Secretariat. The result of a lack of data in order to better identify potentials of 'know-circulate' of the diaspora is one of the major aspects. If the Commonwealth provides an institutional framework to lead a reflection on migratory politics of states, it makes, however, little use of migrants' associations' reservoir of resources and experiences. The relationship between migration and development is not treated in its entire dimension due to this marginalization of one major actor concerned by this question.

> Migration has historically contributed to development and to human condition. The chiefs of government expressed their concerns regarding existing weaknesses at the level of sending, transit and destination countries of immigration. They demand the implementation of mechanisms of protection for all migrants. (In Document of Declaration of Port of Spain, 2009)

The nexus migration-development fails to be treated in a broader dimension due to non-engagement with one missing actor, migrant associations. This absence is the direct consequence of the privileged option by the Ramphal Commission based mostly on experts and political representatives. This choice of the commission, composed of seven representatives from mainly academic backgrounds, is one of its weaknesses. After presenting the Walter Rodney lecture at the University of Warwick (October 2010), the former Jamaican Prime Minister, P. J. Patterson, supervised the Commission Works. Yet, while recognizing the weight of the diaspora, he advocated an institutional approach left to specialists of the migration issues:

> The Ramphal commission formulates recommendations about immigration policies and its implications for the Commonwealth in the next decades. We are aware about the growing presence of the Caribbean communities overseas as a remarkable force to defend the Caribbean interests in the host society. Their relentless activism has allowed for the emergence of group of interests with transnational objectives. Nevertheless, data are lacking to better grasp this phenomenon and our mission to lead an effective politics of Commonwealth states through reports produced by experts. (P. J. Patterson, University of Warwick, 27 October 2010)

On 23 June 2016, the United Kingdom held a referendum on the question 'Should the United Kingdom remain a member of the European Union or leave the European Union?' The referendum resulted in a 72 per cent turnout of eligible voters, with 52 per cent of the voters choosing 'Leave' and 48 per cent choosing 'Remain'. Experts have rarely got everything wrong from start to finish to the point that electoral politics deserve to be rethought. Against this backdrop, the deployment of a discourse of populist crisis to characterize Trump's governance has reached an overwhelming neoliberal and neoconservative backlash in the media. Yet, serious and real crises continue to exist, such as the European Union's fragile Mediterranean border and its trade policy, but also, importantly, the opportunities for cities in the Global South (Beirut, Lebanon) exposed to collective actions by and in support of refugees. Yet, little or non-existent consultation with diaspora associations is noted in spite of their growing engagement with local officials in both host and sending states. This is one of the tasks of the Ramphal Commission, as it focuses on cooperation between sending states and host countries for a mixed management of migratory waves within the Commonwealth members' state. The Commonwealth provides an institutional framework through which to conduct a thorough reflection on migratory policies of the member states. The recent initiative Ramphal report (Laurent et al. 2017) addresses the broader concern in terms of the uncertainties of economic partnerships between the UK (the third major contributor to the European Developmen Fund (EDF) behind Germany and France) the Caribbean, African and Asian countries at large.

Britain, generally thought to be a multi-cultural society, displays a rich variety of family forms, traditions, and close and continuing links with extensive kinship networks originating in, or extending to, a number of other European countries, South and East Asia, Africa, and the Caribbean. The Commonwealth suddenly becomes the one space that brings together different nations with unique histories and diverse values. On the world stage, it can continue to serve as a rallying point instead of the EU entry point it used to be. Despite this magnificent ethnic and cultural diversity, the Commonwealth members are able to have a consensual multicultural vision of policymaking. A certain Anglo-Saxon culture of openness, tolerance and respect for differences has infused the Commonwealth family and enabled people of different cultures and civilizations to live and work with each other. Such a spirit may not suffice in terms of decolonial thinking, which allows for counting on one's own resources first. In this world, a symbol of hope needs to be strengthened and revitalized. The UK, with its unique position within the Commonwealth, can seize such opportunities to enhance its global position and role after Brexit. Moving away from Pax Britannica or the Monroe Doctrine, regional bodies (GCC, BRICS) watching the development of events closely have a chance to strengthen their EU and UK connections and, thus, to restructure relations and outlooks with both powers and the 'superpower' of the United States of America, with its undisputed world leadership. Their accumulated political wisdom and a common neo-colonial memory can best serve South-South cooperation, diplomacy and displaced crisis management as well as global interests such as climate change.

My discussion of Francophonie and Commonwealth is part of a broader geopolitical reflection about the impact of associative action of Haitian and Jamaican residents of both institutional and multinational spaces where they are less visible and under-researched. The future of Francophonie in Haiti, depending on the Haitian diaspora's influence on French territory, is also tied to cultural diversity as a major focus of cooperation among states. The pertinent dialectic between emigration and Francophonie emerges out of a multiple spatially diverse space (francophone Africa, Quebec), producer of immaterial and tangible wealth for the sake of the development of host and origin society. Whether dealing with tensions of integration or state-to-state relationships or multilateral fora (ACP, EU) in a post-Brexit context, it is in the 'dualism which is established between the 'periphery', where the colonized are, and the 'centre' where the colonizers are located (Nafafe 2009) that part of the dilemma is engaged.

Contrasting extra-territorial citizenship: Commemorating Haitian–Jamaican Day

Building on the work of Ma Mung (2004), I favour the geographer's perspective that places much more stress on spatial relationships on different scales, where specific diasporic experiences imply the emergence and the survival of migrant communities. Here, space and place as well as social networks are pertinent to explain how a 'tenth département' is constructed, as part of an extra-territorial unit of the Haitian Republic (Anglade 2002). The community abroad is viewed as an extension of the nation; however, this informal relationship calls into question the legality of citizenship. Wah (2003) suggests the difficulty for dispersed migrant communities in representing a single administrative unit inscribed into Haiti's own geographical map. Active citizenship goes beyond traditional loyalty towards one nation and brings about fluidity in terms of exchanges across space. I draw from the experiences of both the Haitian and the Jamaican communities to think through the ways politics has mediated their relationship with their host society and homeland. This sheds light on the challenges facing the migrant's associative actions in order to find an operational framework adapted to describe what is taking place on the ground.

In deconstructing citizenship, Sassen also argues that the narrow legal definition of citizenship does not adequately explain how citizens within a particular polity, and minority groups, experience the institution differently, though citizens may not benefit from the social and political dividends of citizenship (Sassen 2005).

> Let's take the case of the Haitian consulate in Guadeloupe. The team in charge keeps changing. The consular representation causes problems since its transferring in 1993 from Martinique. In less than two years, the consul was changed due to a chaotic handling of Haitian refugees in Guadeloupe in the aftermath of a military coup between 1991 and 1994. Some consular agents were put aside ('economic licenciés'). These events gave a negative image of the consulate and generated a less positive ambiance. In order to promote development, internal organization needs to be thought seriously. This is the only way to attract investment from the younger generation as well as foreign capital. (Interview with Jojo, consulate of Haiti, Guadeloupe, 29 January 2010)

Going back to the conceptualization of the Haitian diaspora as a 'tenth département' already discussed to the recent evolution of diaspora politics via a *cosmonational* model, the rise of transnational politics remains a challenge to territorially based approaches to citizenship and immigration policies. Michel Laguerre (xiii: 2013) calls cosmonation a 'cross-border cultural set of people claiming the same ancestry ... evolving at multiple diasporic sites'. Although the diaspora is viewed as an extension of the nation, many issues must be resolved before this informal relationship can be placed on a solid legal ground. The metaphor of the collapse of all the disparate emigrant sites into a single administrative unit is inscribed in Haiti's own geographical map, which becomes a social and political map in which Haitians settlements abroad are the main element.

Because various tentacles of the Haitian diaspora are spread across the United States, a central committee was established by Jean-Bertrand Aristide's administration to coordinate the activities of the 'tenth département' to promote the return of Aristide to Port-au-Prince to complete his term in office. A grassroots leader was selected for each region. This structure created several problems for both official representatives and overseas communities. One of the obstacles to the creation of an inclusive and transparent process had to do with the problematic selection of the heads of those 'tenth département' units throughout the diaspora. In the case of France, some interviewees denounced the non-transparency and non-representativeness of some of the appointed leaders who came from other places instead of being part of the local communities. As a result, some conflicts arose in questioning the legitimacy of some official representatives as opposed to grassroots community leaders who had control over a larger constituency. Since some of these local organizations predated the conceptualization of the 'tenth département' as an administrative concept, some resisted cooptation and preferred to continue their traditional linkage to specific regions of Haiti. Initially, it is a political structure attached to the Office of the President that differs from other model of diaspora affairs unit (for example, the case of India) under the Ministry of Foreign Affairs. In 1994 the Ministry of Haitians Living Abroad (MHAVE) was established as the official unit responsible for the coordination of diaspora relations to the government.

The relations of the representatives of the tenth département to the official overseas representatives of the Haitian government (ambassador and consular officers) still need to be clarified. The frequent direct contacts between diaspora leaders and US officials at times undermine the effectiveness of the diplomatic corps. The strength of the diaspora resides in its fundraising capacity. In addition to financing the electoral campaigns in the elections of 1990, it raised a quarter of a million dollars to finance projects of the new government. Its resources can tip the balance toward one side of the political spectrum or the other. Whether the central committee coordinating the activities of the diaspora is seen as an independent voluntary association, an overseas leg of a political party or part of the government in office will determine the level of participation of the people and the survival of the organization.

The 'tenth département' initially used as a political engine to mobilize support for the former Haitian president Jean Bertrand Aristide, became a ministry since 1995 known as the ministry of Haitians Abroad (MHAVE). Yet, after all these years, in spite of the efforts of some representatives, it continues to fall short of its promises of a space in which to rallying progressive talents for the sake of the economic development of Haiti. Public opinion is divided among those who perceive this structure with a section France, with a provisional committee in May 1991 (Fleurimond 2003: 21). Moïse, member of the association Anacaona for the defence of human rights, judges with optimism efforts to unite Haitian communities abroad while voicing some doubts about the way it functions:

> Talking about a tenth *département* appears unfit given the great potential of Haitians abroad, dispersed throughout the planet. Certainly, this mass of people allows those inside the country to live better. Yet, it would be great to show some interest for the protection of their rights. There is a lot to be done on the part of the Haitian authorities to protect Haitian citizens in the Dominican Republic for example, or even illegal in French Guiana or in the Bahamas. If the tenth département is the recognition of our existence and we are given the right to vote, then we will be on the right democratic path after Duvalier regime! We need to look at other examples of countries that have organize their diaspora with recognition of them as full-fledge citizenship with voting rights. How can one invest if not considered as citizen in the country? Right now, I do not entertain any particular return project. (Interview with Moise, retired architect, Paris, 8 May 2008)

The different initiatives at the institutional level are recorded for the period between 1978 and 2012. Table 6.1 gives a contrasting view of how approaching the diaspora may trigger different responses in both Haitian and Jamaican case.

Table 6.1 Organizing a politics of memory through institutional design

Case of Actor	Date	Institutional Set-up	Objectives & Initiatives with Diaspora
Haiti	26/12/1978	Decree of creation of the Organization of the Service of Immigration and Emigration	
	29/3/1987	Vote of the Constitution (24th): Article 15	Double citizenship is not recognized

Case of Actor	Date	Institutional Set-up	Objectives & Initiatives with Diaspora
	16/2/1989	Decree of creation of the Commission of Haitians *d'outre-mer* (COHOM)	
	9/1991	Commissariat of Haïtians of the *Tenth Department*, Haïti (COHDID)	
	25/1/1995	Ministerial law creating the Ministry of Haitians Living Abroad (MHAVE)	
	30/3/1995	Decree creation of Migration National Office	
	16/8/1998	Diaspora week in Haiti	
	02/07/2002	Law granting privileges to citizens of Haitian origin with dual nationality (Loi Voltaire)	
	07/09/2007	Selection of projects dealing with the framework of *codevelopment* SCAC	
	07/09/2010	Project of transfer of human skills and competence (with PNUD & OIM)	Haitian Diaspora Forum Haitian Diaspora Organizations
	21/3/2010	Sponsor Initiative of Organization of American States (OAS), Washington, DC	Participation of three diaspora associations (France, USA, Canada) along with officials
	16/3/2011	Presidential decree official Diaspora Day	
	20/4/2011		'Haiti se nou, nou tout se Haiti' (Haiti is Us, we are all Haiti) (Event, Petionville)

Case of Actor	Date	Institutional Set-up	Objectives & Initiatives with Diaspora
Jamaica			
	1993	Charter for Returning Residents	Protocol for Returning Residents Facilitation Unit (RRFU)
	1998	Overseas Department (JOD)	
	2003	Jamaica & Consular Affairs Department	Series of dialogue (USA, UK, Canada) [NAJASO]
	2004	1st Jamaican Diaspora Conference	Establishment of a Jamaica Diaspora Foundation/ Institute
	2006	2nd Diaspora Conference (Kingston)	1st UK Jamaican Diaspora Conference
	2007	Jamaica Enterprise Network JEN-UK	To encourage business development within Jamaican community in the UK
	2008	3rd Diaspora Conference (Kingston)	
	2011	4th Diaspora Convention (Ocho Rios)	Partnership between Jamaica Stock Exchange (JSE) -Caribbean Enterprise Network CEN-UK
	2012	5th Diaspora Conference (London)	

While conducting the fieldwork, one point of convergence was expressed in the significance of contributing to solidarity funding.

Space creates the possibility of creating a state of reciprocal flux of ideas but also capital and memory. Symbols such as the flag or green landscape can play in the consciousness to trigger sense of belonging and solidarity. Despite some respondents indifferent or negative image of a structure like the ministry of Haitian Abroad (MHAVE) often called 'empty shell', transfer capital or local investment continue to be on the rise. To the question 'what does the MHAVE represent to you?' twenty-three people out of eighty interviewed responded 'not much'. Forty-two individuals considered the role of the ministry in supporting various projects in the country important. Finally, fifteen interviewees had a negative opinion and showed their mistrust in regard to the Haitian institutional structure. Siméus, of the association *Haïti Mémoire et Culture*, reflected on the broad scope of the role of MHAVE:

> At my last voyage in September 2001, I was in particular touched to revisit my family originally from Léogâne even though many reside now in Canada and in the United States. I dream of returning in Haiti with an investment in business; I was hopeful to attend the programme titled 'Vacances pour l'avenir' (Vacations for the future), organised in 1997 for the youth coming from France, New York and Canada, during the first mandate of President Préval. The objective of this programme was to encourage young professionals to return to Haiti, and to facilitate their reinsertion. We attended a seminar at the Université Quisqueya, we met with some ministers, and we visited the North department of the North to appreciate the possibilities at a local level.

An example of state politics of memory towards its diaspora comes back frequently in my interviews with Jamaican public servants and associative actors in Birmingham. This is particularly noticeable when compared to Indian state policy which favours initiatives among the diaspora throughout the world, and in particular to its population which has profitably settled in the United States. Research has demonstrated how the economic and political interests of the Indian state have been boosted through smart and relentless networking among the Indian communities abroad. The idea for entrepreneurs to serve as lobbyists in Washington, DC in both American chambers in favour of India's emerging markets serves as a model for states seeking a pillar to bridge the gap for bilateral trade deals (Therwath 2007: 36).

> Given the importance of the Jamaican diaspora (about 3 million outside) and its contribution as the main source of foreign currency towards the debt, it is perfectly legitimate for the Jamaican government to consider in the long run a ministry solely devoted to diaspora issues. (Interview with Delano Franklyn, former Jamaican Minister of Foreign Affairs and Foreign Trade, Kingston, 22 June 2011)

The Jamaican government, since independence in 1962, has sought to encourage a systematic and coordinated approach in its relation to its overseas communities. Yet, it is only since 2004 that the Jamaican state has been given a new impulse by officially calling for a formal diaspora conference. During an interview with Lucy, who is in charge of the management of the Platform Jamaica Diaspora in the United Kingdom (JDUK), the emergence of this structure was discussed:

> The consular missions continue to play an important role in facilitating the support for overseas communities in developing project. Generally speaking, members of the government while officially visiting a country will take the opportunity to organize meeting with diaspora communities. On January 12, 1993, the Prime-Minister of Jamaica, P. J. Patterson authorized the creation of a unit of facilitation for returnees (Return Residents Facilitation Unit, RRFU). In 1998, the government decided that the RRFU will be upgraded to a department of Jamaicans abroad. (Jamaicans Overseas Department, JOD). (Interview with Lucy, entrepreneur, 5 November 2010)

From the department of Jamaican Abroad (JOD), a systematic approach was adopted to better coordinate relations with overseas communities for better human and financial utilization into the economy. Yet, this structure lacked the material means to fulfil its mission. In November 1998, the issue regarding overseas communities and their protection abroad came to the Senate floor. A motion was discussed and adopted in unanimity for the recognition of the positive role of Jamaicans abroad in the development of the country. It also calls for the launching of a conference and the possible designation of a minister of diaspora affairs.

In November 1998, the issue regarding overseas communities and their protection abroad came on the Senate floor. A motion was largely discussed and adopted in unanimity for the recognition of the positive role of Jamaicans abroad in the development of the country. It also calls for the launching of a conference and the possible designation of a minister of diaspora affairs.

After the first Jamaican diaspora conference in Kingston in 2004, members of the Jamaican delegation residing in the UK met in Birmingham. The main goal was to discuss the type of structure to put in place to ensure the existence of a 'diasporic movement' sufficiently dynamic and capable to forge a diasporic consciousness. The JDUK was conceived as a structure to be present at the regional level in tune with the spread of the Jamaican presence throughout the UK. Depending on the concentration of the community, one or two coordinators were designated. Mel, manager responsible for the implementation of the JDUK and networking, told me about his fieldwork in each region, which gave rise to a tentative design for the JDUK (Table 6.2):

> One of the first challenges was to clearly explain what diaspora meant as for many Jamaicans who consider themselves much more integrated into the British society than entertaining the thought of being in close proximity with the Jamaican state. Issues of suspicion, negative image were commonly voiced; however, I think that now the community is passed that. The structure initially was very simple with no secretary, no treasury nor vice-treasury. We turned towards regions in the UK, for the objective was to extend the participation and to ensure that dispersed communities were engaged. The decision to divide the UK in six regions has allowed for having in each region interest groups on some key issues. (Interview with Mel, Birmingham, manager of JDUK, 28 June 2010)

Table 6.2 Tentative Jamaican Diaspora UK framework

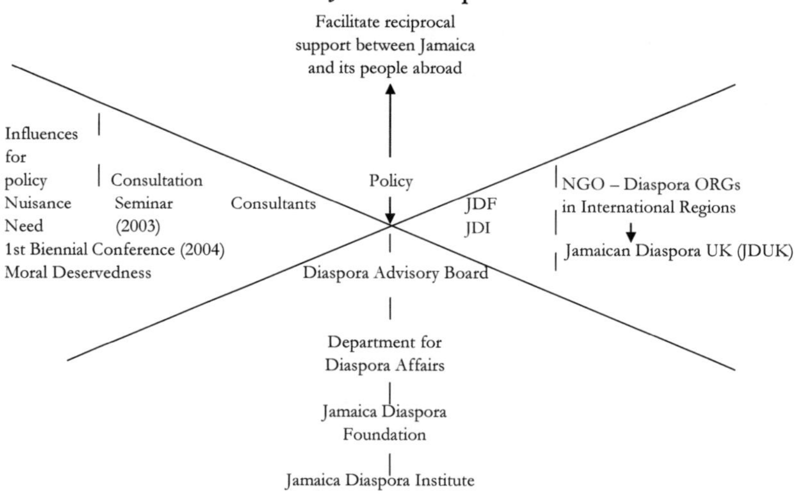

JDUK

1. Organize national entity to represent interests of natural Diaspora to Diaspora Advisory Board – Elect ABM
2. Organize Biennial Conference to : A) celebrate Jamaica Day
3. Bring key institutions from Jamaica to Diaspora (Mgt Board and Regional Co-ordinators)

Source: Template out of group meeting JDUK, Birmingham, June 2011.

Jamaicans in the UK showed a great deal of interest in health and social issues. A group of interest was created to deal with these issues as well as questions of education, arts and culture, and sports. An initiative was taken to discuss issues on church and religious groups, commercial and economic opportunities, rights and political engagement, and finally about public relations and communication. One of the concerns dealt with the image of Jamaica and what the diaspora could do to make it better. At the end of the discussion, a consensus was found about putting together a temporary structure called 'interim structure' which would be subject to later revision.

Commemorating the day of diaspora in the United Kingdom (July 2006) was a milestone in the launching of JDUK. The day was launched under the leadership of the High Commission who greatly contributed with its staff to the realization of this event. This took place despite critics from some associations such as the African Caribbean Self-Help Organization (ACHSO) who were suspicious about a top-down authoritarian style control over the associations. At the very first Jamaican diaspora conference in the UK in April 2006, 400 people were expected. The final count was up to 700, and a sense of satisfaction for the organizers was definitely present. When I asked about the objective of this conference, I was told that the goal was to evaluate the nature of the relationship between the institutional representatives and the associations. Also, what was at stake was the level of interest of Jamaicans abroad in their sense of belonging to the homeland, the motherland, Jamaica.

> Since the National Jamaican Diaspora Conference in 2004, it was hoped to have the constitution of an associative structure managed by an advisory board and serving as intermediary between the consular affairs and the community in the UK. An evaluation phase was urgent to better appreciate to what extent Jamaicans in the UK were interested in engaging in associations with the JDUK initiative. The goal was to dialogue with the greater number, and to do so, it was necessary to come up with adequate tools to inform us about Jamaicans feeling. One of the major challenges was about financing while some felt that it was supposed to mainly come from the ministry of foreign affairs and external trade. (Interview with Jane, member of the advisory board of JDUK, Milton Keys, 8 December 2010)

As a member of the JDUK advisory board, Jane gives an interesting, detailed account of how the dedication of a few actors led to a more significant assembly embodied in the very first Jamaican diaspora conference in the United Kingdom:

> One of the major successes of the programme was the workshop done on immigration issues which reveals the expectation of so many Jamaican citizens leaving in illegal conditions in the United Kingdom. Some of them have neither economic power nor visibility to benefit from legal support. They fear denunciations or expulsions. We anticipated 100 attendees, but finally about 500 people were present. This huge participation reveals the necessity for this type of workshop, and the legal team decided to plan mobile events throughout the country; however, it was very limited. The cost of about £35 000 or the equivalent of 4 million Jamaican dollars to organize this conference was financed mainly by generous sponsors such as Jamaican companies settled in the UK, such as Jamaica National, Victoria Mutual, J. Wray & Nephew and Air Jamaica. (Interview with Jane, member of the JDUK advisory board, Milton Keys, 8 December 2010)

Based on Jane's account, the very first Jamaican Conference of the JDUK held in the United Kingdom went well in spite of some negative aspects. Certain critics denounced the elitist side of this initiative, and in particular the statist weight in the organization of this event. Finally, the approach chosen by the JDUK structure consists in seeking a partnership with individuals and the private sector. As a result of three different meetings (2008, 2010 and 2012), considerable efforts have been made to encourage greater participation of Jamaican associations. The Jamaican community in the UK is older and more established than the Jamaican communities in the US and Canada, which explains the dynamism of the Jamaican associations in British society.

> When we began this work in 2003, we spent lots of time to plan, to document our meeting and in role of secretary of the organizational committee; I witnessed a long and painful process. The problem in this type of conference is that at every meeting groups of individuals are different, apart from a hard core of tireless followers. (Interview with Jane, member of the JDUK, Milton Keys, 8 December 2010)

If several associations have long existed before the rally of 2004 around a structure such as the JDUK, it remains to be seen whether this first initiative will be transformative enough to favour a serious dialogue between the Jamaican state and its oldest community abroad in the UK. A Jamaican university professor based in Birmingham told me what she thought was the best approach and her ideas on the rapprochement done by the Jamaican state. The policy consisted in bringing more cohesion to the multiple Jamaican associations throughout the United Kingdom in a way that might prove profitable to the development of the homeland:

> The choice made by the Jamaican state is in tune with the current approach because I noticed the existence of Indian structures here with their diaspora. If one is able to plan a targeted return of professionals whose expertise is in great need could be the right approach for Jamaica in so many competitive sectors. So, the state must have a clear strategic thinking to launch initiatives towards the diaspora of the United Kingdom. This is why it is called JDUK. At the first conference I attended in 2006 in Birmingham, I was more in the role of an observer since I did not belong to any association. I noticed some enthusiasm on the Jamaican officials' part. Perhaps their interests were not necessarily converging with those of the associations. A coherent angle must be sought to go beyond emotions so the possibility of articulating a viable and inclusive policy can happen. Just by claiming that we are Jamaican means different things for different people. (Interview with Victor, academic, University of Birmingham, 24 October 2010)

This comment of the Jamaican sociologist, Victor, reveals how a country adapts and implements a policy by institutional practices towards its overseas communities. It also underscores the extent to which associations carry specific characteristics to forge a constructive partnership with the state. Since 1998, the Jamaican government integrated in its commercial projections the weight of the remittances. A series of institutional measures were taken to set out a framework to accompany private initiatives from the diaspora. It is interesting to see how a politics of memory is being put in place top down while energizing the base for local engagement commitment. The Trelawny Association (TORA) is one of the Jamaican associations actively working simultaneously from Walsall, UK, to Trelawny.

Remembering the green of Deeside (Trelawny) while living in Walsall (UK)

The relationship between natural catastrophes and the nature of civic engagement from Caribbean migrants' associations allows for more comparative research. The example of the Jamaican case is telling. In 1988, after hurricane Gilbert, Jamaica saw a 2 per cent decline of its GDP in spite of its forecast of growth of 5 per cent for the same year. In this type of catastrophic situation, how is associative mobilization assessed inside the country affected by the disaster and within communities abroad? Sam has lived through this natural catastrophe in his native parish of Trelawny. He decided to stay to assist victims of the hurricane. Trelawny Overseas Relief Association (TORA) came about ten years later as a result of his mobility between New York, Walsall (UK), and Trelawny. He arrived in 1968 as a soldier and later settled in the West Midlands at Walsall. Originally from Deeside (parish of Trelawny), he was first enrolled in the British army and later became a Pentecostal pastor while finding ways to become politically involved in Walsall council. Today Sam is retired and spends time with its association. As the main representative of the JDUK in Walsall, Sam is now focused on political participation. He worked hard to get this association to be recognized as a vector of integration as well as a uniting point within the British community.

The city of Walsall with a population of over 251,000 people (UK Census, 2001) is located in the north of Birmingham. Its municipal council, in a report of 2008, disclosed a relatively strong presence of (14.2 per cent) among staff of people of black minority communities (BME). The scope of those TORA's activities presided over by Sam is a socio-educative association at Walsall, where a community of about 4,000 people of Jamaican origin resides. In March 2007, Sam launched the *Walsall Jamaican House* with the full support of the mayor of Walsall and the High commissioner of Jamaica. Sam maintains cordial ties with different communities and takes the lead in pushing for projects in solidarity with his own community. His engagement draws greatly on the mobility he has enjoyed since his arrival in the United Kingdom and later, in the county of Walsall, where his understanding of the social cohesion policy boosted his socio-political action toward the community at large.

In the case of Walsall, the question is, 'Who benefits from migrant resources, and how do they affect the rules of the political game?' While transnational politics requires a certain degree of mobility and financial as well as human capital, the associational figures presented in this study share two elements. First, they represent a new sense of belonging combined with a form of active citizenship based on actions and not on strict normative criteria. As such, the second feature resides in the overlapping influence in a transnational political field meant to lobby and, at the same time, with some economic constraints. How can we explain such an engagement, which has largely preceded initiatives from the Jamaican authorities through the JDUK?

The fact that Sam managed early on to immerse himself in British society and, in particular, in the county of Walsall, partly explains his active associative engagement. His institutional background in the military paved the way for a smoother immersion later in his civilian life and activism in Walsall's local politics. Sam became the representative of the Jamaican Diaspora UK (JDUK) and, in this capacity, he demonstrated that it was possible to create political linkages with an established framework designed to transcend political boundaries. Beyond spatial mobilities that arise from associational engagement, we must question the true vision of development from the perspectives of people who travel back and forth between Walsall and Deeside. Sam's cousin as a municipal council member, keeps him informed of current discussions on local needs which he in turns relays to the Walsall council. Sam's school project and a storage room for materials during hurricane season are among helpful initiatives for improvements in Trelawny.

Another field in which many returnees to Deeside (Figure 6.1) get involved is the building of new houses. I walked with Sam through several streets and noticed the ongoing production of residential spaces by migrants' investments. This translates into achievement of a long-term goal in the migratory returning phase process. We met returnees such as Sam's sister who decided to trust a local construction company to build her house. The return goes through several phases: from back and forth trips to maintain the ties, then through the commitment to give meaning to the migratory project after years of living abroad, and finally, the need to inscribe initiatives in a much broader, sustainable development scheme at the local level. Under some local mayors' urging, as in Trelawny, more engagement with some associative leaders from the diaspora seems forthcoming, according to Leo from the Association of Returning Residents in Jamaica (ARRR). Leo resumes well Sam's associative engagement in the parish. It is about a way to live *here* and act *there*, by developing associative practices with local officials from the same locality.

Figure 6.1 Maroon land and green of Deeside.
Photo by author: Deeside, Parish of Trelawny (Jamaica, June 2011).

Conclusion

The recent debate about the Windrush debacle reminds of the possible threats that Brexit can pose in a broader political context for non-UK and EU citizens residing in the United Kingdom. Many observers have pointed that the Windrush was not an administrative error, but rather a consequence of longstanding deeply ingrained, but flawed notions and the cultural, political attitudes that were essentially the same that permitted the UK public's acceptance of the injustices of colonialism, imperialism and slavery. Most importantly for the scope of this chapter, understandings and perceptions of sense of place continue to shape the conscious or unconscious policy discourse about citizenship.

At the same time, in multiple ways, new belongings are emerging and call for a critical rethinking intellectual effort to knowledge production. The ultimate goal here is to spark a new type of discourse about how regional governments need to move away from the failed classical development aid frameworks to embrace a paradigm shift that requires decolonial thinking. I contend that a creative dialectic is needed that is based on a set of differentiated development strategies to prioritize migrants' collective remittances, achieve reparatory justice for a particular colonial memory and employ a holistic actor approach to spur local development throughout the Global South. In this respect, the potential of transnational citizenship fueled by a sense of collective memory as a policy lever is explored in the next chapter in terms of a coherent policy for local development.

CHAPTER 7

Rearranging the past to present: The spatial dimension of development

> As long as municipalities fail to value decentralization to unlock local development opportunities while slowing down the control of the central power in Port-au-Prince on resources in the periphery, it will be vain to hope for their success over the struggle for their full autonomy.
>
> — Interview with the Mayor of Abricots, Jean-Claude Fignolé, 2 July 2011. He died 11 July 2017

This chapter contributes to the study of transnational entrepreneurship and politics by challenging its over-emphasis on identities. In contrast, I argue for a deeper engagement of notions of mobility and spatiality at the municipal scale (localities). New mobility patterns and other routes have emerged in localities often situated between borders that have not been fully analysed so far. The case of cross-border mobilization between Brazil, Suriname and French Guiana favours migrants' entrepreneurial investment 'here and there' while reinforcing loyal memories through religious beliefs. Religion as social capital remains an under-researched area of transnational religious dynamics from a comparative perspective. This contribution fills the gap on the significance of networks and mobility in the spiritual experience as a collective development tool and political engagement for Caribbean unity.

(Re)making border mobilities through local partnership and entrepreneurship

Rearranging the past to organize the present, the current research draws on semi-directed interviews with local representatives in Abricots (in southern Haiti) and the Guadeloupean commune of Morne-à-l'Eau to demonstrate the unexplored potential of cooperation among partners of different interests. Solidarity and efforts at reconstruction are expressed through inter-Caribbean partnerships between migrants' associations and mayors from the French region of Oise (Villiers-Saint-Paul and Montataire).

Under the initiative of the Association franco-haïtienne de solidarité et d'échanges culturels (AFHSEC) in Oise, Sheila, organizer of many cultural events, has launched many solidarity actions at the communal level in the south of Haiti (Jacmel, Aux Abricots). After the 2010 earthquake in Haiti, some associations in Oise, such as the Association pour le Développement et la Formation en Haïti (ADFH), based at Villiers-Saint-Paul and led by Martin along with the support of others such as Morse, a retired municipal counsellor of Montataire, joined forces with Sheila to launch, on 16 January 2010, le Collectif Haïti Solidarité Oise (CHASO). This association facilitated the recognition of relations between municipal elected officials of Bretagne along with Côtes d'Armor and the commune Aux Abricots. Before this post-earthquake initiative of CHASO, Sheila, originally from the city of Jacmel (South of Haiti), had been very active in the region of Oise.

Overall, if other associations aim to become viable partners at the communal level, the question is, 'How does associational support become pertinent and efficient?' Certainly, some associative structures may bring an immediate assistance, as is already the case for a humanitarian NGO during the catastrophe. In Sheila's case, the thinking is to go beyond urgency and to inscribe a scheme of solidarity useful to the benefit of the local development. From such a perspective, the challenge comes from a complex process that requires more resources and dialogue with the people on the ground. As a migrant actor, Sheila is linked to a locality through a transnational network which gives her the capacity to use her association in Oise to act as a potential bridge between elected officials in Oise and those in the commune Abricots. To date, cooperation with the municipalities of Oise has made available the funding of approximately 30,000 euros for different types of projects between the two partners.

As I explored the relationship between associations and local officials, the result on the ground was not quite clear. The mayor of Abricots denounced the lack of coordination between spontaneous initiatives and a well-thought-out decentralized development framework at the communal level. He at times even questioned the ambiguous attitude of some Haitians abroad towards engaging on the ground without much discussion and dialogue with local officials. This runs the risk of ultimately having the same negative result observed with some of the NGOs throughout the country. Thus, a more decentralized cooperation, whereby a structure composed of elected officials from *here* and *there* are parallel to an inter-Haitian cooperation among migrants' associations, requires a better long-term perspective based on local demands and needs. In that sense, Sheila and her partners are determined to take such discussion back to Oise.

My findings point to a rapid expansion of associational life. Most interestingly, it stresses the instrumental role of migrant associations in rethinking a form of active citizenship between *here* and *there* based on a sense of belonging to the homeland and to the host country in a decentralized context on a transnational/ local scale. Whether in the Haitian or Jamaican case or in the French/ British context, there is a meaningful evolution in the institutional discourse on migrant civic engagement and local development of their municipalities. What lessons can be learned from the two diasporic models? First, the institutional pattern of homeland-oriented mobilization is central to sustainable sociopolitical and economic development of the Caribbean. Second, since the Haitian earthquake of 2010, the case against aid development has often been argued. Much less attention has been given to the untapped entrepreneurial dimension in transnational communities' contributions to the homeland economy.

Making a living across St-Laurent-du-Maroni to Suriname and Brazil

Understanding Haitian entrepreneurial dynamism requires first contextualizing its relation to the host society along with the neighboring cities. Caribbean workers from all islands have a tradition of staying, both on a voluntary and involuntary basis, in search of jobs and better living conditions. The tobacco fields of Cuba, the building of the Panama Canal, the sugarcane fields of the Dominican Republic: all of this form the backdrop to understanding old and new patterns that continue to feed the consciousness of the Caribbean (Goulbourne, Thomas-Hope, 2009).

Two of the associative leaders I interviewed in French Guiana, Tony and Loulou clearly understood the pertinence and potential gain of adopting an appropriation strategy of the French Guianese space. This requires relying on community resources acquired through activities such as religious assembly or civic mobilization on behalf of 'sans-papiers' (illegal residents). These encounters demonstrated the challenge to the territorially based approach of citizenship and immigration policies as rightly pointed out by Lafleur (2013).

Spatial patterns are vital to the livelihoods of many families in the southern part of Haiti connected to French Guiana and beyond. As already demonstrated in the previous sections, commercial networks increase whenever the mainstream job market remains inaccessible to citizens and non-citizens. As a result, different forms of mobility are constructed to bypass institutional host setting. If long-distance nationalism appears as an idea of belonging that links people who live in different geographic poles and places, it finds its roots in the way people take ownership of themselves with the territory (territoriality) and the notion of transnational solidarity expressed in diverse forms of socio-spatial practices across borders. In so doing, a whole process of deterritorialization brings about multiple loyalties between the state of origin and its citizens living abroad. From this long-distance engagement, dense networks connect migrants with family left behind. Fito, a native of Croix-des-Bouquets (Haiti), is a young entrepreneur living in St-Laurent-du-Maroni, where he launched his accounting company with an associate. Having commercial ties with the city of Albina (Suriname), he regularly crosses the border into French Guiana. He wants to change this 'attitude of isolation' among the Haitian community. He believes that self-confidence is the main challenge to improving his community.

> We have a community which lives for the most part in isolation between parties and religious events; otherwise, nothing exciting and new is happening. The majority of the people work in agriculture banana or sugar cane. It is a source of pride for us to put our energy into an enterprise that has existed since 2009. (Interview with Fito, entrepreneur, 9 March 2010)

Access to money offers these young entrepreneurs a rare opportunity to assemble resources, make behavioural changes, and discourage entrepreneurial dependence. Fritz, a commercial partner of Fito, recounts the beginning of this initiative:

> It was very difficult to convince my entourage and then obtain a contact from a bank before launching the business. Our service is the only type of business of this kind that works with the Haitian community in particular here in St-Laurent-du-Maroni, Iracubo, and Mana. We have a contract with the Consulate of Haiti in French Guiana that allows us to assist with passport applications and birth certificates. I feel completely integrated in the Haitian community. I help to the best of my ability and I consider myself very well immersed in my host community. Our clientele is mixed and of different origins, whether Brazilian, Surinamese or Haitian or Guyanese. We have entrepreneurs working mainly in the construction and agriculture sectors. The great geographic position of Saint-Laurent near the border makes it a very dynamic economic area. (Interview with Fritz, entrepreneur, St-Laurent-du-Maroni, 9 March 2010)

During our visit to the Chamber of Commerce and Arts and Crafts (*Chambre des Métiers et de l'Artisanat*) at Saint-Laurent-du-Maroni, the vice director described the growing involvement of entrepreneurs of Haitian descent. She helped in the procedures necessary to create their businesses.

> We are receiving requests in many sectors, such as restaurants (especially street vending and flea markets) in the food industry (supermarkets). For any profession in the food sector there is mandatory training and, thereafter, they can open their stores close to the schools. We also have files of Haitian entrepreneurs who are dealing with the clothing sector. They participate regularly at our trainings. In the construction business, we observe a positive evolution of their businesses which are growing quickly. For fourteen years, since I have been at the Chambre des Métiers et de l'Artisanat, I can attest to the growing and vibrant activity among people originally from Haiti. They are what I consider good models of integration. (Interview with Mrs Omana Balmin, public officer, St-Laurent-du-Maroni, 9 March 2010)

There are two main factors driving entrepreneurial dynamism in the cross-border mobility from Guiana to Suriname. Nadine, originally from Aquin, is another entrepreneur of Saint-Laurent-du-Maroni who often goes to Kourou. Since the earthquake in Haiti she has belonged to the Confédération Associative de Convergence Humanitaire et d'Echanges (CACHE) and to Collectif Kourou. Her husband works in Suriname.

> For three years, I did all kinds of menial jobs; I opened my accounting business in 2007, named 'Kourou Prestation administratif'. My clients are small construction enterprises. My husband works as an automobile salesman in Albina (Suriname) and helps me replenish my stock in the shop. (Interview with Nadine, business entrepreneur, St Laurent-du-Maroni, 12 March 2010)

As the Haitian presence in Saint-Laurent-du-Maroni keeps growing with cross-border mobilities, entrepreneurs are increasingly investing in the construction business. Henry, originally from Saint-Louis-du-Sud, left Haiti in 1979, but his family and friends remained in the country. He met his future wife, who is also from his hometown, through an association composed mainly of members from his native commune. By virtue of this alliance, his activities diversified from construction to agriculture and cattle farming.

> I am residing in Matoury where I started in the construction business and then I became involved in farming and agriculture. I own land and conduct several farming activities while my wife manages two sheep and beef herds in savane Matiti. She grew up in Saint-Louis-du-Sud (Haitian commune) and knows this domain well. She maintained ties with her parents at Paramaribo (Suriname) and these people regularly replenish her stock. (Interview with Henry, entrepreneur, Eau-Lisette, 11 March 2010)

These different entrepreneurial trajectories shed light on new configurations of 'migratory movement that is neither unidirectional in a bipolar space of departure-arrival, nor of settlement-return. It is rather about a pluri-directional dynamic in relation to diverse spaces' (Berthomière and Hily 2006, 75).

> I have traveled three times already to Haiti since my arrival here: in 2001, 2003, and 2005. It gave me great pleasure to visit my family. I wish to invest in Haiti and to create an orphanage via an association dedicated to social work. The goal of buying property is part of my future projects; the back-and-forth movement fits me perfectly

> as mobility helps me to make ends meet with my family obligations here and there. I send remittances of about 300 euros every three months. (Interview with Nadine, Kourou, 12 March 2010)

Nadine's story shows this young generation's loyalty to their place of origin and their surviving ties to their homeland:

> My mother was already settled in French Guiana when I joined her at the age of fourteen. Unfortunately, she passed away two years after my arrival in 1976. They refused to give me the visa of sojourn and they placed me under social assistance. I married a citizen of Suriname of Haitian origin (Thomazeau). I lost my child three days after his birth. All these sad events would have been worse if it were not for my extended family's support here. I went back to study to become a computer and accounting secretary and I also joined a landscape program and graduated in 2006. I became involved in the associative movement by being a member of the Confédération Associative de Convergence Humanitaire et d'Echanges (CACHE), created in 1991 at Kourou, with the objective of promoting co-operation among different socio-cultural ethnicities and teaching. Up until now, there was no association offering any alphabetization lessons to the Haitian community. We currently have two employees in the association. (Interview with Nadine, Kourou, 12 March 2010)

What is happening in Saint-Laurent-du-Maroni may be put into perspective with the case of two entrepreneurs who settled in Guadeloupe. They supplied their trade using the same cross-border circuits by linking the cities of Suriname with Saint-Laurent-du-Maroni. Rose managed a dress shop in Pétionville before she left Haiti. After her mother's death, she spent some time in Guadeloupe and decided to settle in the city of St-François after meeting her husband.

> When I arrived in 2003, I gathered information about becoming an entrepreneur because my initial project was to start a sewing shop. I found several small fabric businesses managed by Haitian ladies downtown between the streets of Schœlcher, Jean Jaurès, Général de Gaulle, Abbé Grégoire, and Frébault, such as the boutiques Pélican Caraïbéen, Psaumes 21 Shop, PAM Couture, and Christiane Boutique. Such dynamism encouraged me to launch a more ambitious venture with the idea of attracting a chic and wealthy clientele. Victor, one of my friends back in Haiti, has been the owner of more than five boutiques since 1976; he gave me full support by introducing me to Syrian business circles that were quite powerful in downtown Point-à-Pitre. The possibility of maintaining ties opened doors for me. Victor and

> his colleague Fidel taught me about their strategy of cross-border business between Saint-Martin, la Dominique and Guadeloupe in trading practices. (Interview with Rose, entrepreneur, St François, 01 February 2010)

When it comes to entrepreneurship in Saint-Laurent-du-Maroni, social capital is an integrated structure which can move different forms of skills and practices, such as high degree of mutual trust and understanding. This is demonstrated by Nadine and other female entrepreneurs such as Mama in Mana. They supplied their trade using the same cross-border circuits by linking the cities of Suriname with Saint-Laurent-du-Maroni. In this case, connections within the host society in French Guiana helped them identify the opportunity and get into business in other words, made it possible for them to become a part of the transnational network. Viewed in this context, transnational networks allow new economic markets to be conquered; for mobile people, rethinking new paradigms of development becomes essential. From the city of Albina or Paramaribo (Suriname) to the city of Saint-Laurent, from Kourou to the city of Saint-Louis-du-Sud (Haiti), trends indicate an intensification of the Haitian presence, including in Brazil, from Manaus and Tabatinga, where (since the earthquake of 2010) complex networks have quickly been constructed. Indeed, among those arriving in 2010 and 2011, the three primary cities of origin are Gonaives, Ganthier, and Croix-des-bouquets; in 2012, many others came from Port-au-Prince. This is largely a post-earthquake phenomenon, which contrasts with a much older migration towards French Guiana or even Venezuela, which was known as a prime destination for political refugees during the Duvalier era. In any case, the same strategy exists without having to necessarily leave the host country. It is possible to engage in a cycle of back and forth movements that allows for both exploiting economic opportunities here (host society) and satisfying demands over there (homeland), using ethnic networks. These are essential supports to nurture commercial practices that evolve around merchandise for a specific culture. Imported products are designed to supply small businesses and to benefit a friend or family member.

From a spiritual standpoint, the Haitian community of French Guiana appropriates itself the religious living space to reconnect through a common sense of belonging. Such belonging may take various forms, depending on the nature of the religious belief. Such a process gives birth to a transformation: it takes root in a unique proximity through living memory between the Haitian citizens 'here' with the territory and in the relevancy of associative action from a transnational to a local scale. The support of the church is already organized (Pentecostal or Catholic) and grows through networks to reach a much larger part of the community that has not necessarily found their place in the host country. Nevertheless, these communities have the capacity to build transnational networks while staying engaged in the local Guyanese environment.

Through religious belonging (Protestant/Pentecostal or Catholic), I examine the weight of the sacred within the associative dynamics of promoting economic development and social justice in the state of origin. In particular, I question in particular, the role of the spiritual network as a social capital via the individuals' participation in Christian and Pentecostal churches in French Guiana. A spiritual network includes the totality of ties maintained by individuals who belong to religious groups. To understand how religion is experienced from a distance on Haitian or Jamaican soil, it is worth mentioning the characteristics of transnational religious organizations as 'negotiated' with decentralized networks (Levitt 2004: 7). It is within the dominant sense of belonging to the locality of origin that Pastor Paulin explains the expansion of community churches ('communautés-églises'):

> It is after a dense network of ties emerged in Cayenne, then later in Guadeloupe, out of the region of Aquin (south of Haiti) that we observed an enormous flow of Haitian migrants settling in the Hexagon. The members of the Protestant community were coming from the most part from the same families, often in the same rural municipality. (Interview with Paulin, Paris, 26 May 2008)

Whether it is in Mana, Kourou, or the Parisian banlieu of Essone, Haitian evangelical churches are spreading everywhere in the Guyanese environment. Our aim is to understand how migrants have, in such spaces, transplanted a familiar form of group organization (i.e. belonging expressed through a set of initiatives of collective interest) for the benefit of the south of Haiti in particular. In March 2010, I established some contacts in Kourou, Cité Brutus, and Saint-Laurent-du Maroni in preparation for my field trip. This experience confirmed the character of a certain uniqueness (*unicité*) of the Haitian associative space as it has thus far developed on French soil, meaning mostly through family and relational networks giving rise to a case of interpolarity (Table 7.1).

Table 7.1 Repairing broken links through interpolarity

Indicators	Essone	French Guiana
Total population	1,205,850	274,652
Immigrant population	93,500	46,576
Haitian population	1,069	14,143**
% of the immigrant population	4.09%	30.37%
No. of Haitian associations identified	14	26
Main origin cities	Aquin (La Colline ; Fond-des-Blancs) ; Nippes ; Léôgane	Aquin (La Colline ; Fond-des-Blancs ; St-Louis-du-Sud ; Zanglais Nippes (Miragoâne, Fond-des-Nègres)
No. of churches identified	5	13

Source: Author; INSEE census 2014.

Starting first in Cayenne, we observed the existence of evangelical churches in cities with a strong Haitian presence. For example, in Cabassou, we located four Haitian churches: Eglise Baptiste de Cabassou, Eglise Evangélique Baptiste Primitive de Guyane, Temple Pentecôte Croisade Evangélique des Prêcheurs d'Hommes, and Eglise Néo-Apostolique. At Eau Lisette, we identified two assemblies: l'Assemblée Evangélique de la Grâce and l'Assemblée Nazaréenne de Cayenne. A native of Anse-à-Veau (Department of Nippes), Pastor Colin arrived in French Guiana in 1982; he belonged to the Protestant denomination of his locality:

> I arrived alone leaving my four children and my wife in Haiti. In the beginning, things were very tough ... living in a precarious neighborhood at Cité Eau Lisette. My support for the radio station Mozaik is quite evident because this team helped all of us to rally and to discuss the problems of the community. This is a space that united our people and even beyond the Haitian community. I do not belong to any association. My sense of belonging is lived through my loyalty to the Church, Eglise

> Nazaréenne de Cayenne, which feels like a big family to me. (Interview with Pastor Colin, Cayenne, 7 March 2010)

The Assemblée Nazaréenne de Cayenne, located near a densely populated Haitian enclave, attracts many believers at Eau Lisette. After finishing his pastoral studies, Colin became the chief pastor at the Eglise Baptiste de Cabassou, which maintains strong ties with the Baptist mission of Bethesda that is present in Haiti and Paris' XVIIIème *arrondissement*. Since his last trip to Haiti in 2006, Colin has stayed in contact with his family in La Colline:

> Our assembly manages an education centre and we may obtain, after four years of study, a diploma of pastoral study as a predicator. Likewise, we encourage the pastoral care of migrants with other churches in order to work toward the regularization of *sans-papiers*. Our prayers for our land of Haiti aim for internal mobilization in the hearts of people with the understanding of making a better vision of the country and improving the mission of each and every one of us. (Interview with Colin, pastor, Cayenne, 7 March 2010)

The Protestant belonging of Haitian migrants in French Guiana is complex; it involves a double sense of belonging to the community of believers and the homeland. Table 7.1 reveals the links between churches in Haiti and French Guiana. Pastor Brice, the oldest Haitian worker at the distillery of Saint-Maurice, talks about his sense of belonging to his hometown:

> I came here in 1979, running away from the mountain erosion at La Colline. I joined my father who had been settled since 1976 at Saint-Laurent-du-Maroni. The sugar cane industry was booming and I have stayed until now. I took back the church created by my father, *l'Eglise Combattant de la Croix*. Every August, I bring a group of believers to La Colline to reconnect with our village and to preserve the living memory of the mother church. (Interview with Brice, Cayenne, 7 March 2010)

Apart from the protestant movement, there is also an interesting dynamic in the city of Mana; indeed, founded by Sister Anne-Marie Javouhey as a farming colony in 1828, the commune of Mana has since preserved the statute of this emblematic Catholic figure at the entrance of the city. Built in July 1840, the parochial church of St Joseph de Cluny dominates the city of Mana, which is located 40 kilometres to the north-east of Saint-Laurent-du-Maroni (the nearest large city); it has 7,900 habitants (2006). At the parish of Saint-Joseph de Mana in Mana, once a month the choir holds a mass in Haitian Creole for the Haitian community, which has approximately 100 Catholics.

Box 7.1 Resisting invisibility: A catholic 'soul' in Mana

Jacob, a native of St Louis du Sud (Haiti), remains attached to his family and participates, from a distance, in projects for the benefit of the Union des Saint Louisiens de France pour le développement de St-Louis-du-Sud (USLOAFRADES). Jacob settled in Cayenne in French Guiana in 1982, and later in Mana, where he found employment at an agricultural farm. 'These were the good years when there was lots of work in the rice fields. Today, it is much more difficult to find a job'. Since the end of the 1970s, Jacob has never stopped being a fervent Catholic volunteer in the grass-roots movement of 'Ti Legliz' in Haiti. The central presence of the Church of Saint-Joseph de Cluny reminds him of his cousin who was educated at the Sister of St Joseph at Cayes-Jacmel (South-East of Haiti). 'I would like to be everywhere where there is good to do', the founder of this congregation liked to say. 'At Mana, we continue to have our Catholic faith as in Haiti. Father Michel and the Church of Saint Joseph bring us human warmth that is rare to find in the Guianese creoles'. The community organizes itself to pull together around activities with St Joseph Church; it is very useful for the cohesion of a community of Christians. Jacob is particularly active within the church choir and directs songs and psalms in Haitian Creole. 'Since the earthquake in Haiti, many activities of solidarity at the Church and among the Guyanese and Haitian associations have occurred across Guiana. Here at Mana, we chose to rally our forces mainly in prayer. We have organized fundraising and collected clothing to benefit the poor people of Haiti, via the Collectif pour Haiti en Guyane, created on January 26th, 2010. A bigger coordination of initiatives is desirable in the future because the distances are long and the communication is not always easy. I talked to Célio from the association of USLOAFRADES at Sarcelles and my Association Solidarité des communautés de Guyane to coordinate action in our locality of origin. The challenge will be to mobilize our meagre resources to ensure the direct sending of aid, because the State in the capital may forget one more time the people of St Louisiens'.

Source: Interview with Jacob, farmer, Mana, 10 March 2010.

Rethinking voting as an act of remembrance of ongoing chaos, debt and quakes

Studies of memory frequently discuss the role of the state in shaping citizen behaviour towards elections. Yet, little is known about how immigrants' political engagement is shaped by their memory and political loyalties to the homeland. In contrast to Tony or Loulou's political investment in French Guiana, the American political landscape has a long history of immigrants' political and civic struggle. The steady growth of Haitian presence in the United States since the 1950s provides an opportunity to question how influential it has been in this recent 2016 presidential election. While concerns remain about the loyalties of 'hyphenated Americans', the widespread acceptance of multiculturalism in American society has legitimized activities by ethnic groups to advocate within the US political system on behalf of their country of origin and its interest. My primary aim was to assess how the political engagement of Haitian-Americans affected the outcome of the 2016 US election and whether their influence on US foreign policy towards the homeland could make a difference in the future.

Following the threat of Hurricane Matthew in the southern part of Haiti during the night of 3 to 4 October 2016, the US presidential debate brought the issue of Haiti and the troubling role of the Clinton Foundation to the forefront (Corsi 2016: 183–221; Seintenfus 2015: 231). This was the first time the issue had been raised in this election. The role of the Clinton Foundation in this event was controversial, to say the least. Hurricane Matthew became the first Category 4 storm to make landfall in Haiti in more than fifty years (National Hurricane Center, South Florida). The storm devastated Haiti, leaving some villages underwater. Yet, the humanitarian neo-liberal rhetoric along with increasing external interference in Haitian matters has rarely been addressed in academic terms. In fact, Haitians abroad have sought to influence politics in Haiti, despite its quite limited success in lobbying the US, French, and other governments. This discussion opens up new questions in our understanding of ideas of citizenship are played out across networks that transcend national borders. A vote 'here' carries meaning 'there'. In other words, the level of support given to several rallies to denounce one specific candidate has more to do with Haiti than US politics on domestic front. I argue that the decentralized nature of the American political system allows for transnational communities with limited visibility to maximize their level of civic engagement to exert more influence over both homeland politics and in the US.

Where in the United States of America do Haitian-Americans mostly live? To begin to explore the political participation of voters of Haitian origin, one needs to locate where their presence is spread across the counties of the state of Florida. My visit to the Institute for the Study of International Migration (ISIM) at Georgetown University (January 2017) gave me access to interesting reports and statistics specifically related to the evolution of the Haitian-American community. Many were initially living up north in the states of New York or Massachusetts. Relaxations in the USA's and Canada's restrictive immigration policies caused a rapid growth in immigrant flows by the end of the 1960s. Haitians found themselves in metropolises with a strong dispersion based on solid family networks. For example, their combined arrivals rose to approximately 1,500 people by 1966. An estimated 350,000 to 400,000 Haitians live in the south Florida area. They make up 4.2 per cent of the population of Miami-Dade County, currently the highest concentration in the US (FAVACA 2008).

In contrast, the situation of the Haitian-born community in French Guiana paints a very tough picture, especially at times of social movements such as the one that has been active since March 2017. This turmoil has coincided with the presidential election campaigns. Numerous protests in Kourou, the symbolically significant site of the European Union-financed base Centre Spatial Guyanais are keeping the tension high. A group known as the Collectif 500 Frères, supported by thirty-seven unions, is leading this general strike, and has called for the immediate arrival of the French Prime Minister Bernard Cazeneuve. They are calling for 2.5 billion euro, and a national development plan for Guiana. This conflict reveals the weakness of the institutional structure of overseas *départements*. A survey of the National Institute of Statistics (INSEE) 2012 publication reveals numerous socio-economic indicators reflective of many economic weaknesses compared to the rest of France, and even its other overseas *départments*. In 2012 the per capita gross domestic product in French Guiana was at 15,416 euro. This is half that of the Hexagon

(31,420 euro) based on data presented in a recent parliamentary report by the current Guadeloupean deputy-mayor. These statistics serve as an introductory background to the critical situation faced by a more vulnerable segment of the population: the immigrants. Against this critical socio-economic background, I will discuss the strong Haitian presence in French Guiana as a dynamic critical mass favourable to political engagement in various elections. The significant progression of this participation finds its explanation first within the activism of a group of political opponents to the Duvalier regime in Haiti. Many of them transited here in the mid-1970s while others settled later in Venezuela. They were Marxists, social-democrats or Christian-democrats, working for human and political rights across different cities in French Guiana.

Box 7.2 Memory of the struggle

'We left Haiti but Haiti is still with us fifty years later! Our engagement within the Haitian diaspora in the land of Félix Eboué is kept alive'. This moving statement, was made by Yves André Rameau, president of the Association Konbit, during a conference on 25 January 2014 at the *mairie* of Matoury. It is the starting point of a wider testimony to the linkages that exist between the Haitian community, its roots, and the host Guyanese society. Professor Milot Oxigène, originally from Haiti, arrived in Guiana aged eleven in 1966. He noticed that there was more frequent debate about Haitian communities settled in North America (United States and Canada) mainly due to its more significant demographic presence. 'The former deputy of Guiana and History research fellow, Paulin-Christian Bruné, stated that between 1963 and 2013, the community became diversified in its composition. Indeed, he noted that the first Haitian immigrants came barefoot, with no education but with the firm intention of sending their children to school. 'In spite of their illiteracy, they worked very hard to ensure their kids remained in school. After obtaining and education equivalent to that of most Guyanese, these children chose to come back to Guiana. Today, many of them have interesting posts here, and some are involved in politics, such as in the *mairie* of Kourou, of Matoury or of Cayenne. Others are at the forefront of the struggle against xenophobia and precarious living conditions.'

Between associative activism in the socio-educational domain and political engagement, transnational practices were observed in the field during the March 2010 regional elections in French Guiana. During the political campaign in Haiti in 2010, the president of the General Council of French Guiana met with one of the candidates, Mirlande Hyppolite Manigat, General Secretary of the Rassemblement des Démocrates Nationaux Progressistes d'Haïti (RDNP). Among this political party's supporters are those such as Tony and Jeanjean in Cayenne. Both took a prominent role in organizing meetings with the candidate. We met with Tony, with whom we discussed the evolution of the Haitian political engagement and its implication for both Haiti and French Guiana:

> If a political party wishes to have Haitians in its list, it needs to go through the bureau of ASCOMA-RDNP. We choose a person capable of being eligible and not just a figure head. This is how we understand our work of lobbying in order to promote Haitian active participation in the political system of French Guiana. For me, the Haitian community is integrated once all Haitians are accepted just like the Chinese, the St Lucians or others. Where do you find a Haitian assistant-mayor, or in what medical office do you find a Haitian nurse? To talk about integration, you need to have Haitians better represented at all levels of the society. We do have a growing, positive group of students in many of the best schools. At the same time, at the space station in Kourou, there are a few Haitian engineers. Recognition of our rights and our integration must go through elections and active political participation. Parties that need our vote must put us on their electoral list. (Interview with Tony, cité Brutus, 6 March 2010)

Tony's statements are part of an ongoing debate about the shortcomings of French citizenship, particularly in overseas territories. He is a naturalized citizen whose demands and expectations are shared widely among the Haitian community. Many engaged in associations seek recognition of a double belonging while working to serve the collective interest of the society. Regarding this form of civic participation made of multiple allegiances, the current debate about the relationship between immigration and citizenship shows the degree to which political incorporation is influenced by the politics of naturalization (Brubaker 1992: 24). Regarding the result of the regional elections of March 2010 in French Guiana, Loulou sees it as 'a true test' for the Haitian community in finding its key role in the political landscape:

> The Haitian vote usually tends to be in favour of the left party; however, this time it is going to be a pragmatic and useful vote given to the right with the list of *Guyane 73*. We are sensitive to those who give us a hand. The mayor of Cayenne has recently appointed one of our compatriots to the municipality to an important post. He teaches in a school in Balata in the vocational section. We are proud of this professional achievement, and we want to support him as an example for the younger generation. With one or more representatives of Haitian origin within the council, we will be better able to send a positive image of our community and its overall potential in French Guiana. It will help us attain some claims: a better programme to integrate asylum seekers, and a transparent process for getting a residence permit. There is a long list of demands to the public Guyanese authorities who act as if French Guiana was an isolated fortress. It is a big country with lots of resources, and it is under-populated with approximately 230,000 people. (Political meeting with Loulou, Rémire-Monjoly, 11 March 2010)

One element of the media that promotes the socio-political aims and discourse of the Haitian community in French Guiana is the radio station Mosaïque, located at the heart of cité Brutus (Cayenne). It is one of the pillars of the Haitian community and has three purposes that reinforce the role of the associations. Radio Mosaïque is the only Haitian radio managed by Haitians in a French overseas territory. It is a symbol in itself and can be compared with radio stations in cities (Miami, Fort-Lauderdale, Montréal ...). It then expanded into multiple connections with radio stations across Haiti and neighbouring Latin American countries, where a web of complex networks serves many already settled and others searching for new opportunities across borders. Radio Mosaïque serves as a conduit to inform or rally electoral choices. It serves as a socializing space to several generations of Haitians, offering a diverse range of programmes (information, culture, sport or even entertainment). Several interviews alluded to emotional recollections through folk music heard on the radio. For the younger generation like Al, the process of memory is in a constant state of renegotiation in interaction with the local music influences. Al, twenty-three years old, born in Port-au-Prince, arrived in Guiana in 2002. He only obtained legal residential status in 2010. His volunteering to participate in a morning radio show gave him the opportunity to discover and to develop his talent as an entertainer. Through the socio-educative activism of the association AIDE, he was given school assistance.

In 2001, only one candidate of Haitian origin participated in the municipal elections in Rémire-Montjoly. Six years later, the action of some associative leaders became more visible during a rally, the 'Congrès des naturalisés (the assembly of naturalized citizens), which resulted in the election of one Haitian in the main city of Cayenne. In 2010, three further Haitians were elected at the municipal council of Kourou and two in Matoury (Calmont 2012: 13). Loulou's remarks during the last electoral meeting of March 2010 reveal a turning point for the political consciousness of the Haitian community:

> Our community has long been stigmatized. The time has come to mobilise for serious political integration! Yet, we have demonstrated our civic loyalty to French Guiana through socio-educational actions with the association AIDE. During this campaign, radio Mosaïque welcomed men and women from all walks of life to discuss their position. We want that our community just like the others, to be respected and recognised for its work for French Guiana. (Loulou, candidate, political meeting, Cité Brutus, March 2010)

Box 7.3 Faith-based activists in DC, Maryland and Virginia

During my fieldwork in the different areas of DC, Maryland and Virginia I visited a number of associations, either affiliated with various religious denominations or not. Given that Maryland has the biggest Haitian presence, the bulk of my interviews and meetings were concentrated in Maryland. Regarding the church activity in this area between Washington, DC, Maryland and Virginia, Reverend Constant noted that several Catholic churches throughout Maryland and also in Washington, DC offered services in Creole. 'Eglise Sainte Trinité is Episcopalian and very progressive. There is not much difference between the one in Haiti and here. Pastor Holly came in 1851 from Connecticut, New Heaven, as part of a global movement of Afro-Americans who settled in Liberia and others in Haiti.' Then I discussed with Julie and Teresa, two members of another church very active since their years in King's County (Brooklyn) in community education. In South Florida, they have relatives engaged in faith-based circles close to the US Republican party. Their association Clinica Michaa is a base to connect between Miami, their home in Mirebalais and here.

Source: Interview with Reverend Jean, 16 January 2017, Maryland.

Conclusion

This chapter has attempted to draw together Guianese and American cases as illustrations of three categories of transnational political activities as demonstrated by Lafleur (2005: 27). First, homeland politics covers political activities exercised with the host society by overseas communities in specific domains linked to the homeland. These can work to either support or to express hostility towards the political regime in power at home. Second, immigrant politics are political activities undertaken by migrants in a host society to improve their own living conditions. Finally, translocal politics go beyond government initiatives by directly engaging localities where they originate from (commune, rural towns). Based on this classification, the activities of members of associations (Tony, Loulou or Reverend Jean, Julie) overlap at different moments of their respective journeys as a reflection of three transnational areas. By mobilizing since the 1970s against the Duvalier regime in Haiti, their political action was then turned towards the homeland (Homeland politics). Their activism in the 1980s focused more on the struggle for the 'sans-papiers' and refugees (boat-people in Miami), in the educational domain and for the respect of political rights in the host society (Immigrant politics). Third, their community engagement with officials gave them the legitimacy to become actors to the benefit of their localities of origin (Translocal politics). The next chapter will expand more on how memory intersects with civic engagement transcending the three categories conceptualized by Lafleur. Memory offers ways of drawing together the local and the transnational, the political and the global to reflect new forms of citizenship of great value to geography and environmental sciences.

CHAPTER 8

Narratives of remembrance and Caribbean lessons in post-earthquake Haiti

The role of memory politics is explored through activism marked by remembrance. This book has argued that memory is crucial to understanding transnational citizenship and its impact on strategies of development. Geographical understandings of memory have become crucial to unsettling the discourse on flows of people, capital and ideas that circulate across and through different national spaces. To understand the sense of belonging shared by those with a common colonial past in the Caribbean demands that we revisit the historical roots of the origin state itself. Michel Rolph Trouillot (1995) advocates a view of the history of Haiti as the long tradition of an authoritarian state and has repeatedly pointed to the silences around Haitian history. However, the foundation of Haiti on 1 January 1804 reinforces the particularities of the role of a new state in the game of international relations. None of the many uprisings that punctuated the long struggle against slavery and colonization had achieved such a result. Nor does any other revolution consisted of so many radical elements to provoke a will to silence it at any cost. With this in mind, I will reflect on narratives of remembrance by interviewees, and also consider King Christophe's progressive vision of Haiti, as a means of engaging in a final discussion on envisioning the Caribbean future in the wake of Haiti earthquake in 2010.

Remembrance of the flourishing kingdom of Henry Christophe and his beloved queen

> In less than thirteen years, in a kingdom of less than 14000km², Christophe led the building of nine palaces and fifteen castles, including the well-known Palais de Sans-Souci and Belle-Rivière on the banks of L'Artibonite. Prosperity came from the high volume of commercial exchanges. For the only port of Cap-Haïtien in 1817, 150 boats transported tons of coffee, sugar, cacao, cotton and other goods.
>
> — Jean-Price Mars, 1933

Few academic works have seriously addressed the progressive dimension of the only Haitian monarchy, that of Henry Christophe, other than to focus on the supposed 'despotic character of the ruler' (de Cauna 1977; Crusol 2007). I argue that, by revisiting the journey of King Henry Christophe, the historical context in which he ruled, and his geostrategic alliances with abolitionists, one is able not only to demonstrate the progressive aspects of his leadership, but also the lessons his 'developmentalist' path has for present-day Haiti.

My attendance of the Caribbean Studies Association (CSA) in Grenada in June 2013, thirty years after the American invasion of Haiti in 1983, gave me a rare opportunity to discover Henry Christophe's land of birth. I returned on September 2016 to continue my research on the Caribs, in particular the memory of the Kalinago people, the last Amerindian inhabitants of Grenada. I also explored here, the many parallels and divergences in memories of the Franco-British rivalry as manifested in Grenadian and Haitian consciousness, such as the remnants of Creole language, Catholicism, and a sense of national pride. This research involved several visits to the Grenada National Museum (created in 1976), and a tour of the island's major historical and archaeological sites (parishes of St Patrick, of St Mark). A mobile and multi-sited ethnography offers insights 'into a multitude of mobile, material, embodied practices of making distinctions, relations and places (Büscher et al., 2010 105).

Very few American historians mention the Haitian participation in the Battle of Savannah. Howard Zinn is one of the few who stresses the pertinence of the researcher's positionality. Indeed, I share his principle, as spelt out in his autobiography: 'You can't be neutral on a moving train ... not just better informed, but more prepared to relinquish the safety of silence'. Perhaps the reason for this silence is due to the profound defeat of the allied Franco-American troops by the British. Other historians documented that approximately 800 'Haïtiens libres' called Chasseurs-volontaires, the first such group to appear in the annals of the French armed forces, came to Savannah in September of 1779 with the troops of Comte d'Estaing. This troop managed to reduce damages in this famous battle. D'Estaing, coming from Cap-Français (Haiti), made a stop in St George's and Fort Royal (Grenada), and this is where Henry Christophe (later King of Haiti) was recruited on the way to Savannah. He was born on 6 October 1767, of 'parents libres'. In cooperation with the American army, the unit laid siege to British-occupied Savannah in the colony of Georgia. After many days of bombardment, as well as a failed frontal assault on 9 October 1779, the Americans marched north and the French withdrew to their ships. The same archives (Saint-Rémy 1956) reveal that the corps under the leadership of Comte d'Estaing sailed back to Saint-Domingue. Christophe went to work as a cook at the hotel La Couronne in Cap-Français. This property belonged to Marie-Louise Melgrin Coidavid's family. Born in 1778 on Habitation Bredou, a commune of the cross-border city of Ouanaminthe, she was raised in a family of free coloured ('gens libres') and learned to read and to write. She married Henry Christophe on 15 July 1793. In a surprising twist, I found traces of her later in 2016 in Pisa, Italy.

From its beginning, the northern state of Haiti departed with the Constitution of 1807 through its republican orientation. Four years later, a new constitution was proclaimed by Henry Christophe. This instituted a monarchy and implemented a rigorous policy of economic recovery based on systematic production of major export commodities. This model of organization follows the one used by Toussaint Louverture and Dessalines. Christophe's experience went beyond his ruling of thirteen years of existence, demonstrable even today with substantive material evidence. Given the profound regime crisis of the republic in the west of Haiti under president Alexandre Petion (1807–1818), it is interesting to discuss two major aspects of this unique political experiment in the north of Haiti: an impressive educational system and a complex set of geostrategic alliances. In his correspondence with Wilberforce, King Henry argues for the necessity of cultivating a set of geostrategic alliances that would take Haiti out of the ambit of colonial France. Unlike other generals in the west and the south attached to a republican regime, Christophe organized a royal court through diverse networks that allowed him to constitute a centre of power with a great deal of discipline and respect for protocols. It also allowed him to begin to implement his vision of a productive socio-economic independent position independent of the colonizer, France.

Christophe developed a set of guidelines in the 'Code Henry,' composed of 133 articles, with a civil and penal law, civil and penal procedures, and laws on trade, military service and military criminal law (Leconte 2004: 398–399). Throughout the colonial period, very little formal education was available in Saint Domingue. These letters are vital primary sources that can redefine the king's and the abolitionists' motives regarding their alliances. In a letter to Clarkson in 1819, Christophe asked the abolitionist to hire two professors for the Royal College of Haiti, because the king believed that, due to Wilberforce's illness, Clarkson was 'the friend best placed to understand our needs in the field of education.' The king asked Clarkson to enlist teachers willing to move to Haiti for seven years, insisting that they 'will be honourably treated, and will enjoy in Haiti all the consideration which their profession merits'. Following Wilberforce's advice, Christophe allowed British missionaries to open schools in Haiti during his reign, and he invited two American women to open a public school for girls. This was by no means the first time that missionaries had come to the New World, and indeed the king himself had previously welcomed them to his country; but it was the first example of a ruler in the New World intentionally bringing in missionaries to improve educational and 'moral' conditions of his country. Creating an educated society of former slaves was not enough for the Haitian king: he wanted to prove that education could create racial equality. The king used his reforms to refute racist notions that people of African descent were inherently ineducable and to provide the world with an example of a progressive black nation. The letters between Henry Christophe and the abolitionists (William Wilberforce and Thomas Clarkson) allow this dimension of Christophe, understood as a nuanced and intelligent leader, to be better appreciated. Specifically, the political dynamics between France, England and Haiti in the post-revolutionary period provide a significant lesson for generations to come.

With this palace and the whole of city of Cap-Français having been torched during the Haitian Revolution, Christophe now King demonstrated his sense of grandiosity with the magnificent construction of the Citadelle. It is the most convincing example of Haitian national pride aggravated by the obstacles it faced in its creation. Spread over 8,000 square metres at the top of the Bonnet-à-l'Evêque, 1,000 metres above sea level, and fitted with more than 200 artillery guns capable of defending four cardinal points, the Citadelle resembles a Herculean task forged by a mythological giant. The monument speaks volumes about the profound and obstinate determination of Henry's thirst for grandeur and for the salvation of his people. The magnitude of the Citadelle's design far exceeds the simple goal of defending against French invaders. Centuries have gone by and the monument remains as a temple intended to be more than a symbol of material or military prowess, but also as a lightning rod with which to revive the memory of heroic battles won for many generations of Haitians to come. Citadelle Henri encapsulates the burning desire to keep alive the inviolable right of a people to live free or die.

King Henry managed to leave money in the coffers of La Citadelle in contrast with the financial mismanagement of his rivals in the west and in the south. In addition, his correspondence reveals that one of his most important goals was obtaining international recognition. The German historiography sheds light on a possible correspondence between Henri Christophe and Friedrich Wilhelm III of Prussia, who ruled between 1807 and 1820 after Frederick the Great (1712–1786). He was known for his reformist agenda, marked by political, socio-economic and cultural modernity. It is also documented that the Code Henry (1812) found some inspiration from Berlin and Göttingen. It is also possible that King Henry Christophe used German architectural source to build the Palais of Sans-Souci (Hector 2012).

The marginalization of France by the United States of America in the management of the earthquake of 12 January 2010 confirms the decline of the French influence in Haiti since the end of the American occupation of 1915–1934. This is the case despite there being so many factors favourable to the maintenance of privileged, cordial, and respectful relations between the two states. This state of affairs is explained in part by King Henry having resisted many arrogant claims from French representatives to compensate French plantation owners. In reality, France has never recovered from the loss of its former colony, which was the richest in the world at the time of the French Revolution in 1789. Haiti had 793 sugar plantations, 3,151 indigo plantations, 3,117 coffee plantations, 789 cotton plantations, sixty-nine cacao plantations, 182 rhum and tafia plantations, and many more resources (ISPAN/UNESCO, 1991). Economic revival took place during the thirteen years of Christophe ruling in the North (Figure 8.1). This raises the question of the relative weakness of postcolonial studies in French academia when compared to the British system, and the many silences around the Haitian Revolution that persist. It also militates for the necessity of a different perspective, one involving decolonial thinking, that attempts to deconstruct the official narrative of the content of 'L'histoire de France'.

Figure 8.1 Recollection of the Kingdom of the North.
Source: Author's family collection.

The proclamation of its independence on 1 January 1804 made Haiti the second independent state to have gained independence from a colonial hegemon, following the United States of America. The fraternal assistance given to Simon Bolivar in 1816 by Haiti's president, Alexandre Petion, for the emancipation of the Spanish colonies and the abolition of slavery in Latin America, makes Haiti one of the pioneers of the ideal of Pan-Americanism. Haiti has consistently made an important contribution to the cause of inter-American solidarity. This is why it believes itself to be held in the esteem of all its sisters of America; and this is why it counts on the diplomatic support of its neighbouring countries to facilitate its political, economic and financial development. The recent positioning of Haiti in support of the sovereignty of Venezuela within the Organisation of American States (OAS) can be partly explained by the historical depth of Haiti's relations with other independence movements in the Americas.

In revisiting the sole king in Haiti's history, my aim has been twofold. First, to do justice to an under-researched historical period, and in so doing to challenge the often simplistic and unfair judgement passed down upon Henry Christophe. And secondly, I have sought to appreciate the difficult task launched by a visionary leader in the context of a very dark and inhumane era. The King of England, George III, sent a representative to attend Henry's coronation on 2 June 1811. Two months before, he had created a nobility: four princes, eight dukes, twenty-two counts, thirty-seven barons, and fourteen chevaliers of Saint Henry. Perhaps Aimé Césaire's drama *La tragédie du Roi Christophe* obscured more than it revealed about the true nature of Christophe's kingdom and its legacy. After all, Césaire, during his long presence at the French National Assembly, never claimed to favour an independent path for former colonies. The neo-colonial project, applied in the west and the south of Haiti under the leadership of Petion and later throughout the country since Boyer, sought to mask or even to distort in the collective memory the real significance and the impact of other visions of Haiti, and in particular the one defended by Henri Christophe. Hence, one can note the silence around the work of intellectuals such as Baron Pompée Valentin de Vastey (1781 Ennery) in 'Le Système colonial dévoilé' (1814).

De Vastey's report of 1819 on the educational progress of the kingdom mentioned five schools in which 536 students had received training from British schoolmasters in Port-de-Paix, Saint-Marc, Cap-Henry and Sans-Souci. The idea was to give moral as well as intellectual education. Yet, the silence about the positive aspects of Henri's administration contributed to a longstanding distortion in the interpretation of the 'path implemented in the Kingdom of North for the construction of a nation' (Hector 2009: 255–286).

On 7 July 2011, I visited Cap-Haïtien (former Cap-Henry) the historical heart of the Kingdom of Henri Christophe (Figure 9.1). It was moving to pass the 'Barrière Bouteille' to continue the trip to Limbé where I discovered the library, 'La Bibliothèque Georges Castera', a warm space open to knowledge. Put simply, it felt like a living space for the elevation of the souls! Indeed, what is the Georges Castera Library at Limbé, intended for if not revival, echoing the slogan on the coat of arms of the once famous King Henry I: 'I am reborn of my ashes' (*Je renais de mes cendres*). Indeed, the young generation of Limbé, part of the 'Kingdom of the North', continues to preserve the memory of the mystical blazon of Henry. The red and black flag bears the phoenix crowned with gold decorated with stars, supported by two lions and a royal slogan: 'God, my cause and my sword' (*Dieu, ma cause et mon épée*).

The importance of localizing memory to places such as Fort-de-Joux (France) where Toussaint Louverture spent his final days connects with the multitude of memory narratives that permeate my journey as a researcher. The final leg of my trip, to Italy in September 2016, took me to Pisa, at the Capella of San Donnino, where lies the marble grave of Christophe's beloved Queen Marie-Louise Melgrin Coidavid. She married the sole king of Haiti whose short life ended on 8 October 1820. In the face of hardships and political instability in the north, she was helped by British abolitionists such as Wilberforce to make the journey to London. She arrived there in 1821 with her daughters Améthyste and Athénaïre. She then made it to Pisa, Italy, via the port of Arno located on the Mediterranean coast. Padre Terenzio of the Capuchin Order recounts Queen Marie-Louise's religious devotion and knowledge of Latin, which can be traced back to her pious upbringing in the

Melgrin-Coidavid family in L'Habitation Bédou. This site is part of the communal section of Savane-Longue in the commune of Ouanaminthe (Casséus 2016: 110,178). He also discussed scarce records at the Florence Library which detail her journey from North London to Salzburg, then from Rome to Pisa. Queen Marie-Louise buried Francesca Améthiste in 1831 and Anna Athénaïre in 1839. She herself died on 14 March 1851. On her grave 'A Qui Riposano M. Luisa Vedova De Enrico I Morto Re Di Haiti'[1] (Figures 8.2, 8.3, 8.4 and 8.5).

Figure 8.2 Capella San Donnino, Pisa.
Photo by author: Pisa, Italy (2016).

1 At Capella San Donnino, Pisa, Father Terenzio acts as a gatekeeper of the site where 'rest the mortal remains of Marie-Louise widow of Henry I, deceased King of Haiti'. Legrace Benson (2014: 93) traces what may have been a captivating metamorphosis of Queen Marie-Louise as 'always a Capoise [who] become Haitian, become queen, and finally become an exotic Capoise Haitian commoner located in Pisa'.

Figure 8.3 Faithful gatekeeper.
Photo by author: Pisa, Italy (2016).

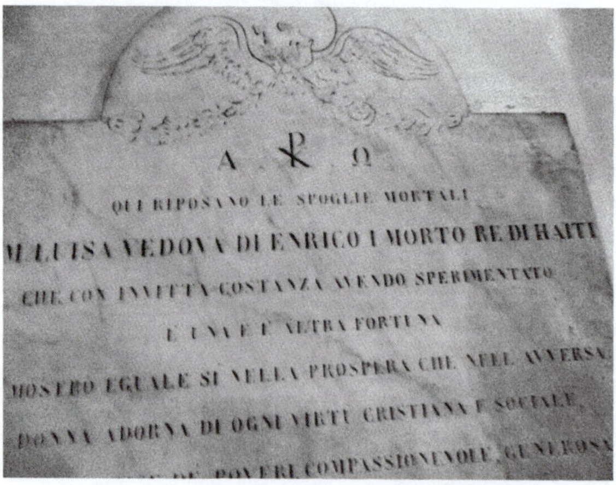

Figure 8.4 Tracing Queen Marie-Louise.
Photo by author: Pisa, Italy (2016).

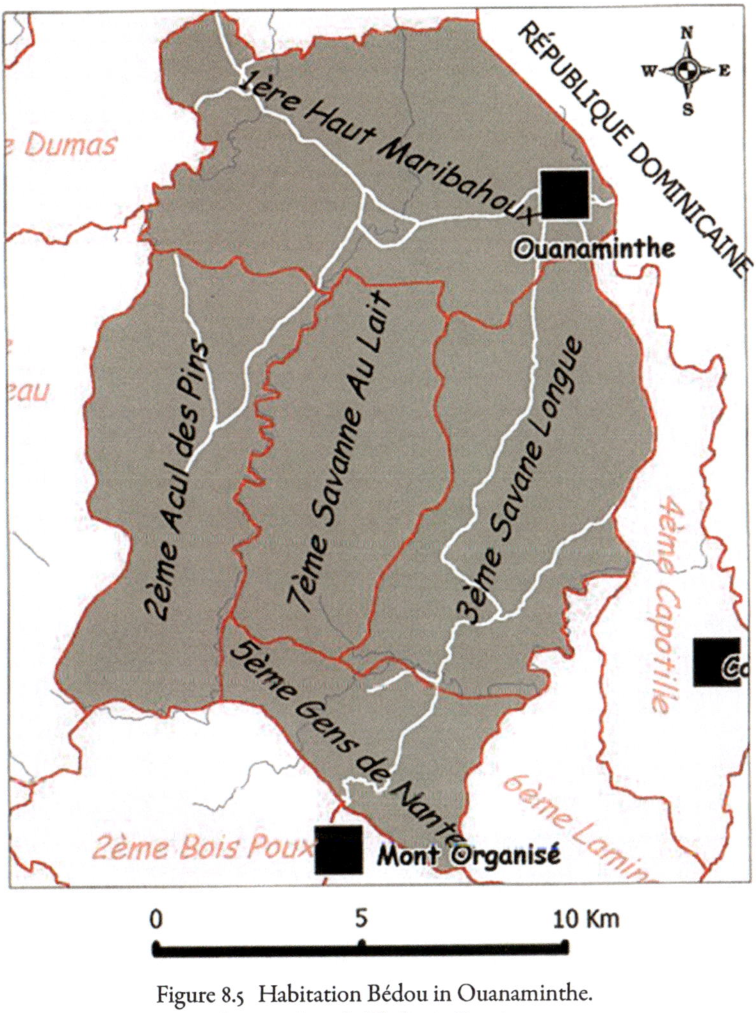

Figure 8.5 Habitation Bédou in Ouanaminthe.
Source: Joseph Vladimir Casséus.

Henry Christophe was one of the rare leaders in the early stages of Haiti who chose a model for development influenced by progressive, Anglican Britain. His determination is expressed early on. In a letter dated 2 February 1802, he declared to General Charles Victor Emmanuel Leclerc, Napoleon's brother-in-law sent to crush the Haitian revolution: 'the land will burn before the fleet anchors off the harbour'. Haiti had the important role of being the first nation ruled by emancipated slaves in an era of racism and slavery. Through his achievements, King Henri I proved that it was possible to be born in a remote Caribbean island, Grenada, to fight in the battle of Savannah for American liberation, and to become a leader capable of achieving in one generation what it had taken Europe centuries to achieve.

Remembering the blue and red flag in London

Throughout the Caribbean, maroons engaged in illicit activities with the broader colonial society. In the specific case of Haiti, and to a lesser extent in Jamaica, the impact of marronage on a culture of resistance and a quest for freedom justifies the widespread practice of runaway. The tales of various female warriors recounted in Chapter 2 continue to reinforce the consciousness of communities abroad and dispersed throughout the diaspora. What are new are the constant narratives of remembrance collected throughout my research whether with respondents of Haitian or Jamaican ancestry. Commemorating Jamaican Day or the Haitian Flag Day cements material or emotional solidarities. After the earthquake of 2010, I encountered interviewees such as Claire, Annette and Mimi who demonstrated a vivid memory and a profound sense of belonging to the homeland. This was reflected interestingly around the celebration of the Flag Day.

> At the beginning, we had small events just to socialize among ourselves, such as Haitian traditional meals and parties. Our activities got bigger and we became a larger group at the 18 May Haitian flag day. Since 1986, after living in the United States, I had no idea of a Haitian presence or community here. It wasn't until 2008, during an encounter with Mia, that I became aware of such a presence. Mia and her husband are founding members of the Haitian Support Group (HSG), an association working towards Haiti's social progress. Her grandfather lived in Haiti in the 1940s, where he worked in rural education. In 2008, after meeting with two other Haitians Sauveur

> and Claire, I decided to join the association United Haitians UK. (Interview with Mimi, London, 21 June 2010)

The flag in the Haitian revolutionary period symbolized the exaltation of certain higher values conceived for a nation that aspired to greatness as a collective body. It was also a utopia with an idealistic meaning that inspired righteous behaviour. Those whose boldness of spirit found both slavery and suicide intolerable would flee to the woods and mountains and form bands of free men. They were called maroons. They fortified their positions with palisades and ditches. Women followed them. They reproduced. And for a hundred years before 1789 the maroons were a source of danger to the colony.

> We are in small numbers living in the United Kingdom, just about 250 people compared to the much greater presence of Haitians in France or elsewhere in Europe. Most of us reside in London, except for a few living in Birmingham, Oxford, Manchester and Southampton. The vast majority of us were born in Haiti even if we left it very young and feel Haitian. We feel grateful to the homeland and Flag Day is one of these days to revive our memory of our land. The goal of our association, United Haitian UK (UHUK), gives us a space to share and to keep the deep sense of belonging with Haiti alive in spite of the long distance. (Interview with Claire, London, 25 June 2010)

In December of 2010, the West Midlands newspaper *The Voice* published an article about Anette, one of those many voices advocating for inter-Caribbean solidarity. Her engagement towards Haiti and her activism within the Afro-Caribbean community to finance the initiative *Haitian Appeal* is striking. Such networks of members of diverse migrant' communities of Caribbean origin throughout the United Kingdom are where Haitian associative dynamism finds its roots and continues to evolve in spite of its limited presence here compared to France. The ties maintained with the homeland are made of utilitarian solidarities and/or emotional memory that feed into a different type of civic engagement. A certain accumulation of resources (human, social and financial capital) through multiple practices are evidence of the nature of long-distance engagement, as shown in several interviews. This element is essential in the analysis of this study, as it confirms the claim that 'the specificity of Great Britain is to possess a deep migrant associative fabric. It is the product of over forty years of race relation' (Lacroix 2009: 31).

Given that these migrants' associative initiatives operate both on an urgent, emergency basis as well as a more long-term orientation, it is worth underlining some legitimate concerns about the chaotic multiplicity of actions labelled 'humanitarian' through NGOs, which are, at times, disconnected from the reality on the ground. Countries such as Haiti have become the target of opportunistic foreign capital in search of big profits and golden business deals. Such concerns are in line with a broader analysis dating back to 1982, which at that time found that 'private investments can not only become illusions for a great number of needy people but also pave the way to a calculated ideological, economic, political and financial tutelage at a national scale' (Anglade 1982: 67).

I respond to this challenge by attempting to conceive of a more complex mode of understanding that avoids eliminating all uncertainty and insufficiency as it involves knowledge of the global. What gets counted as memory and how people involved are doing what they count as 'memory' is of particular interest in migration studies which focus on an interdependent system made of a multiplicity of transnational exchanges between host, origin states and immigrants' associations. By treating simultaneously two geopolitical entities La Francophonie and the Commonwealth, the aim is to avoid reductionism (blind to holistic qualities), unilateralism (the choice to take a part of the whole), holism (blind to relations in parts), and finally to go beyond the common obstacles of a limited and fragmented knowledge. My final claim questions the correct roadmap to seek best reparation and 'repair' using the CARICOM institutional framework.

Debt and the Haitian quake: Repairing the past through CARICOM framework

> The French refused to recognise Haiti's independence and declared it an illegal pariah state. The Americans, whom the Haitians looked to in solidarity as their mentor in independence, refused to recognise them, and offered solidarity instead to the French. The British, who were negotiating with the French to obtain the ownership title to

> Haiti, also moved in solidarity, as did every other nation-state in the Western world. Haiti was isolated at birth – ostracised and denied access to world trade, finance, and institutional development. It was the most vicious example of national strangulation recorded in modern history.
> — Beckles, 2010

Commemorations of the tragic Haitian earthquake give a particular resonance to the observations of the Caribbean Vice-Chancellor of the University of West Indies, Sir Hilary Beckles, quoted above. Many were not aware of this historical event, which is still intentionally silenced in many academic circles. His positioning reminded me of the late Professor Nettleford during the two-day international symposium on the remembrance of the slave trade, organized by UNESCO in 2004: 'The newly independent United States denied to the newly independent Haiti the recognition it earned and deserved to have on the diplomatic international stage; and for the reason that the United States, for all its declaration of inalienable rights and equality from birth of all men, was a slave-owning society'. Of particular pertinence for Caribbean leadership becomes the production of collective memory, understood as 'a political process that is constantly being re-negotiated that 'works out between active agents who remember together' (Halbwachs 1992). In November 2006, former Jamaican prime minister Patterson, declared of the issue of reparation that: 'It is a matter to consider in its entirety. It needs to be better pursuit as a CARICOM initiative because we are in the same boat'. For Shepherd, 'Repatriation is a part of reparation. This discourse must be located into the discourse of development. With unequal international relations, co-operation of proximity is vital for all the Caribbean nations' (*The Price of Memory* 2004).

> Haiti should not have offered tax concessions to the British in 1814. It should not have submitted to the French terms for recognition in 1825. It should not have accepted the terms to reschedule the double debt in 1838. It should not have taken out the loans in 1874, 1875, 1896 and 1910 on the terms that it did ... Haiti was undone by a combination of imperialist intrigue by the United States, abusive behaviour by other members of the core and an economic model that was too dependent on coffee and custom duties. (Bulmer-Thomas 2012: 192)

The question of 'restitution and reparation', already discussed at the Durban Conference of 2001 (Robinson 2013: 246–263) continued to pollute the political life of the country on the eve of the bicentennial of Haiti's independence, which was one of a kind in the history of humanity. Under the direction of professor Harold Toussaint, a group of students in the Social Communication Department of the faculty of the Human Sciences analysed the meaning of the Debray Report for them as students today and for future generations (*Haïti et la France (2003–2004) un débat escamoté* 2009). They concluded that it avoided the real debate about historical facts and the necessity to establish a repayment schedule.

'Is memory the history of the poor?' was the initial question of a pioneering research study of colonization in France (Dewitte 2000). This is at the core of the historical litigation between France and Haiti that keeps coming back. In particular, on 7 April 2003, during the commemoration of the bicentennial anniversary of the death of Toussaint Louverture, a specific royal ordinance of 17 April 1825 raised serious debates and tensions. It stipulates the following:

> The current inhabitants of the French part of *Saint-Domingue* will remit money to the *caisse fédérale des dépôts et consignations de France*, in five equal instalments, year after year, the first due on 1 December 1825, the sum of 150,000,000 Francs, which is requested to compensate former slave owners who demand reparation. We concede, under such circumstances, by the present ordinance, to the inhabitants of the French part of *Saint-Domingue*, the full and entire independence of their government.

During the nineteenth century, France made itself almost the exclusive creditor of Haiti, as a result of this colonial debt 'agreed' and passed on by successive borrowing. French diplomacy in Haiti was based on the theorem of external debt (Pean 2000: 523).

Haiti signed the 1825 Ordinance under the explicit threat of French guns. A naval blockade and the implicit threat of re-conquest and re-enslavement loomed large over the conclusion of the agreement. In fact, this colonial annuity would go to feed financial speculations and to raise the accumulation of capital in France. This point was rarely underscored even when Haiti had fallen for this vicious trap and accepted the unfair terms of the ordinance, the Independence Debt and the loan terms extended by the French central bank. Under 'procedural unconscionability', this could serve as a compelling case to prove that Haiti agreed to pay the French with its back to the wall. For a nation that prides itself on the celebration of modern diplomacy, and with some of its intellectuals addressing Haiti's plight in terms of the 'destiny and curse of underdevelopment' (Blancpain 2001; Debray 2003), France today faces growing criticisms by the CARICOM Reparations Commission (CRC) of the CARICOM Ten Points Plan for Reparatory Justice. Since the earthquake of 2010, much has been written about Haitian resilience when faced with French malevolence. Haiti's place in global geopolitics is neither about a curse nor a destiny, it is about a memory and a history that deserve to dialogue with courage. When the Representative of French Guiana, Christiane Taubira, submitted a first version of the law that bears her name (Taubira Law), it included a clause making reference to the creation of a committee for the purpose of determining prejudice and conditions for reparations. This clause later mysteriously disappeared. In 2002, she intervened in the French National Assembly:

> The debt of colonisation paid to France is not the sole cause of Haiti lagging behind economically and socially. Yet, it unquestionably represents a considerable financial tax, a handicap that limited capital accumulation and modernisation of the productive system while contributing more to the accumulation of capital in Europe than to the most productive colony of the world, called the pearl of Antilles and responsible for such enrichment since the 18th century.

The critical, if not catastrophic, situations, debt management has caused in some Caribbean nations (Jamaica, Haiti, or Puerto Rico) gives a sense of the critical usage of memory and its impact on the motivation of dispersed communities. This thorny question of indebtedness affects more than a dozen Caribbean nations and represents one of the factors explaining the vulnerability of their economies, which are too dependent on remittances (Skelton 2004: 44). The second factor in inequality has to do with the weight of remittances, which are subject to the sole will of the individual or a group of migrants.

> Regarding remittances, it is impossible to impose withholding a percentage for a development fund; this even goes against Jamaican culture, which considers money

> with a watchful eye and the way to spend it as well. In so doing, institutional structures will have to be convincing to make some changes. (Interview with Anthony, High Commissioner to the Court of St James, Ocho Rios, June 2011)

Whether reparation can solve the under-development that still plagues the least-developed country of the Americas, Haiti, or begin to repay the abyssal Jamaican debt, it is difficult to envision the Caribbean future without some types of institutional solidarity mechanism for sustainable development. All fifteen member-governments have signed up to the first-ever regional Reparations Commission for Caribbean Slavery and Native Genocide. The issue of reparatory justice calls for serious research away from electoral manipulations. In the Haitian case, development remains a frequent concern for associative members, and one much discussed in their meetings. The dialogue between MHAVE and their associations are too dependent on political will at home, instead of serious leadership to address considerations regarding a sustainable national development set of strategies. Few skilled migrants contemplate the project of returning let alone investing. In spite of that, there is a general sense of reviving memory through the act of constituting associations across France and its territories. In the years ahead, it will be interesting to map out sites of memories and to connect them with specific Caribbean communities settled in various European cities.

The debate on the true impact of Caricom's decisions and the question of reparation remains an open one. Indeed, this question previously discussed (p. 10) was debated at Durban Conference in 2001 and continued to 'pollute' political life in Haiti immediately before the bicentennial anniversary of its independence in January 2004. New modes of cooperation are on the rise, in particular between states of the Global South, from Cuba to Brazil, from Brazil to Saudi Arabia, from Qatar to Suriname. This means the issue of reparation can no longer be hidden or unilaterally dismissed by French scholars or policymakers. It needs to be seen in terms of a significant economic loss, as well as a moral prejudice affecting past and present generations. The magnitude 7.3 earthquake of 2010 that hit the Western Department of Haiti particularly forcefully re-opened this 'western hole' ('*trou occidental*', Steinfus 2014: 33).

Eight years after the earthquake in Haiti (2010–2018), just over a hundred years since the US occupation of Haiti (1915–2015), two factors bring to light the complex circulatory dynamics across the region: the overlapping of multiple external interferences; and the plantation complex in the mercantile system that the Haitian revolution did not totally dismantle. Indeed, while Latin America and the Caribbean, as well as sub-Saharan Africa, remain among the most unequal regions in terms of revenues sharing, extra-Caribbean migration rose to a peak between 1980 and 1995 in Trinidad and Tobago, the Dominican Republic, Grenada, Guyana, Jamaica and Haiti. These last two countries were particularly affected by migratory flows outside the Caribbean, with 100,000 and 140,000 workers respectively in 2005. Based on these data, we can conclude that the frequent practice of mobility is a constitutive element of Caribbean society. During my interviews, associative and institutional actors shared the view of a positive contribution from migrants in the homeland. Nevertheless, it is questionable to what extent one can appreciate their true impact on the functioning of the society with which they continue to maintain ties for a number of reasons that can be unclear or ambiguous.

In spite of historical and cultural links existing between Haiti and France, Haitian presence in France (the Hexagon and DFA) is difficult to characterize in terms of a 'Haitian community'. In the French Republican context, in spite of a dense associational movement, the visibility of such presence remains marginalized.

The brain drain will likely continue to preoccupy both Haitian and Jamaican policymakers. It will be vital, however, for the government to look at the role of immigration and return in filling skills gaps, as well as the emigration flows that create them. It will also be important to carefully identify the extent to which shortages of healthcare workers are due to decreases in training budgets. Problems of insecurity linked to expulsions or extraditions remain a serious issue that prevent programmes of insertion for the returnees. Going forward means tackling the problem of entrenched organized criminal groups in Jamaica (to which deported criminals often belong). Migrant associations are seeking a better positioning to play the role of bridges and even lobbies. Finally, the approach chosen by the JDUK structure consists in seeking a partnership with individuals and the private sector. Considerable efforts have been made to get a greater participation of Jamaican associations based on three different meetings (2008, 2010 and 2012). The old settlement of Jamaicans in the United Kingdom compared to the one in Canada and the United States of America reveals itself to be one of the explanatory factors to better understand the dynamism of the Jamaican associative life in British society. It is a very complex movement with interesting inter-generational socializing practices (such as a domino tournament throughout the UK, for example) that needs further academic scrutiny post-Brexit.

A long history of indifference between Paris and Port-au-Prince can be understood as if the Haitian revolution had never seriously counted in terms of Haiti's diplomatic relationship with its former hegemon. From the proclamation of independence in 1804 until 1838, the date of the official recognition of the new state, different French governments, under pressure from the remnants of slave-owners' lobby known as Club de Massiac, threatened new expeditions to attempt to put Haitians back in captivity. The defeat of the Napoleonic troops by an army composed mainly of slaves and Creoles was received in Europe as a slap in the face to a group of men who believed themselves to be superior and who were invested in the idea of colonialism as a mission to civilize humanity.

Nevertheless, during the recent French presidential election campaign, Emmanuel Macron, in an interview with the press in Algeria, declared colonialism 'a crime against humanity'. His statement, which caused controversy especially from the far-right and right parties, speaks volumes about what continues to be perceived as a conflicting memory of the same historical fact. Building a memorial of slavery like the one in Guadeloupe (2015) is not enough to reconcile one shared memory. Césaire, in his *Discourse on Colonialism*, equated colonialism with a 'disease ruthlessly gnawing at his country, like a strange virus with no antidote'. Will Macron, the youngest person ever to be elected French president, now call for the recognition of France's past colonial wrongs with a more inclusive history syllabus in schools? Will post-Brexit academic research lead to more effective mapping of the entangled genealogies of capital, debt and quakes in cities overwhelmed by the issue of displacement? These interrogations along with other concerns discussed throughout this book will be part of the concluding chapter.

Conclusion

The aim of this book has been to develop the idea of memory and its implicit relationship with history, places, practices of citizenship and state politics. The notion of the past extends well beyond the way in which power is usually discussed in terms of norms and not necessarily how it might be challenged. This conclusion is arranged into two linked discussions that review implications for envisioning a Caribbean future impacted by new forms of socio-political engagement that may be more important to younger generations of Caribbean heritage residing abroad.

From coloniality of power to the liberation of knowledge

> Can we think about freedom and the future while remaining in colonial jail? We will always be colonised as long as our relationship to history itself is not decolonised.
>
> — Kamel Daoud, 2017

In this conclusion I confront the rethinking of memory with postcolonial discourses and an emerging decolonial epistemology. Over the past several decades, postcolonial geographers have repeatedly reminded the discipline of its colonial past. However, a decolonial geography would be much more focused on how that colonial past is still active in the inequalities of the present and an uneven development calling for the possibilities for reparatory justice. Walter Mignolo's (2010) influential version of decolonial theory proceeds from the setting out of a 'colonial matrix of power' that began under formal imperialism and continues to the present day. It consists of the effective control, by the West: of subjectivities (through religion, and I would argue also through education; see also Tolia-Kelly 2017); of authority or political power (through local and international law-making and policymaking); of the economy (through exploitation of labour and resources); and lastly of knowledge (through control of institutions and technologies of knowledge, such as universities, the printing press and, latterly, the Internet) (Mignolo 2010, 119).

Memory does not have to be the history of the poor. Its role coupled with spiritual beliefs and religious practices, reveals a dynamic and fluid space for solidarity. For expatriate Haitian and Jamaican communities, religion continues to play a fundamental role in the unification of the structure of the community. The old settlement of churches has accompanied successive migratory waves in the British space, hence earning a sort of legitimacy and authority. Pastor Vlad, who was invited to the Jamaica Diaspora Conference in the United Kingdom, declared: 'The survival of the Jamaican community and its contribution to Jamaica depend on the political and socio-spiritual engagement nurtured by the work of the Black Church in England' (London, 16 June 2012). In other words, we are dealing with a spatial and spiritual space, coupled with transnational economic support upon which transnational networks are resting and are able to ensure a civic and political participation in the host country and homeland.

The challenge throughout this book has been to articulate the collective and individual memories of a colonial past as well as the ongoing engagement of communities abroad in order to preserve ties with the homeland. The invocation of both place and flow should alert us as geographers to the ways in which the discipline is not only able to effectively critique the ways in which the discipline is now inserted into ongoing dynamics of coloniality but is also well placed to respond to the call to decolonize knowledge (Noxolo 2017). The question is to speak to the pertinence of decolonial knowledge and how it matters in understanding the culture of the 'other'. One of the fundamental questions confronting tourism geographers is that of 'how places develop as centres of tourism.' More recent understandings of place have explained how patterns of coloniality inform the continuity of the historical burden. By reading globalectically, I am developing in the last implication discussed below a challenging interdisciplinary path informed by Glissant's poetics of relation and Morin's complex thought between geography and politics in relation to migration and critical memory studies.

New geographies of memory: Circularity, heritage tourism and globalectics

Geography and politics have always been closely allied. They were twin disciplines in the ancient cosmological tradition 'that imagined the orderly workings of the geocentric universe as machina mundi, the machine of the world' (Guardiola-Rivera 2010: 58). My argument is that investigating the concrete networks of families of Haitian and Jamaican origin exposes the continuum of colonial legacies at the heart of collective struggles of workers, the displaced, and refugees in urban and rural contexts. A place of settlement is not an important point of reference as a social group with a common national origin develops a strong feeling of attachment to the 'homeland' (Hidle 2001: 2). Local identities are caught up in a web of global interdependence, and circularity helps better understand how national identities are created and formed through empowerment built on resistance to oppressive external forces. What is changing is the geography through which power operates. Hence, an intellectual effort requires that we 'de-transcendentalize the nation and shore up the redistributive powers of the regionalist state in the face of global priorities' (Spivak 2010).

Among the reasons why heritage has been widely adopted as a component of urban regeneration and identity claims are its intersections with collective memory that evolves through time; and its stimulation of socio-political and economic activities fundamental to the emergence of many new forms of belonging. Central to this process is the use of the past as discussed throughout the book in bringing Haiti and Jamaica into a single analytical geopolitical framework. The stories told here – from cross-border entrepreneurial dynamism to voting activism and solidarity funds – demonstrate that the limits of knowledge production encapsulated in sophisticated narratives catering to a primarily literary elitist audience, need to be countered with bottom-up alternative methodological approaches. A deeper understanding of the evolution of the politics of homeland governments which once considered emigres to have abandoned their countries, becomes pertinent in a broader developmental perspective fostered by extra-territorial citizenship embodied in a plurality of engagements.

In fact, more of these sending states today have recognized that migration confers benefits to local economies (such as Trelawny in Jamaica or Abricots in Haiti) through a socio-spatial process shaped by a complex creolized web of transnational practices. Caribbean-born scholars Stuart Hall, C. L. R. James, Edouard Glissant, Rex Nettleford and Nobel laureate Sir Arthur Lewis, have all been instrumental in conceptualizing and defining the specific historical and institutional ground-breaking creative Caribbean space. Some have warned about the risk of replicating global colonial patterns in local settings and doing away with the hierarchical relations of knowledge (Maldonado-Torres 2012). For Dalphinis (2018), Caribbeans have not emerged in high number with creole as part of their agency of social change.

'We are now national urban space, globalised urban scape, and two generations now of virtual space.' Lewis's (2011) statement in paying tribute to the late Professor Rex Nettleford's intuitive and profound knowledge of the Caribbean captures the evolving character of the field of cultural studies that touches on different disciplines. There is an influential relationship between history, sense of belonging and memory that helps to reshape geographies of contemporary politics. Less clear, however, are the multiple means by which religious engagement such as the belonging in faith-based circles (case of Panama City, FL) have truly affected political directions and influenced elections whether at home or in the host society. Reconciling the Caribbean with a past of dispossession, a present of uncertain geopolitical configurations (post-Fidel case), yet a future full of hopes (i.e. the role of *Créolité, Françité and Hispanité* as three vectors of pluri-versality of the Caribbean culture) calls for courageous dialogue on common and conflictive memories.

I therefore return to globalectics, which 'embraces wholeness, interconnectedness, equality of potentiality of parts, tension, and motion. It is a way of thinking and relating to the world. It articulates 'a web of connections of mutual dependence' (Ayo 2017). Power is involved in who gets to represent whom. The museum acknowledges its own role in telling biased stories, as a result of its historical collection development policies during the colonial era. The decolonization of memory starts with a redefinition of what memory is and why it matters in the narratives of Caribbean historiography. Circularity allows rethinking the religious and cultural practices of the colonizers less often analysed in their originated port cities such as La Rochelle, Bordeaux or Nantes. Beyond the ongoing debate within the CARICOM, what is required is a serious historical, educational and cultural work of deconstruction/rehabilitation that involves a sincere transatlantic partnership to learn the lessons of the mistakes and crimes of the past. The search for justice, and the reparation of Caribbean societies that continue to suffer the legacies of slavery and native genocide, was at the heart of the discussions at the second CARICOM Reparation Conference (12 October 2014). The opening of the University of West Indies (UWI, Mona Campus, Jamaica) Reparations Research Centre in October 2017 further indicates the desire to further investigate the past and to challenge the silences around the transatlantic slave trade. The centre has recently called for justice on behalf of the 'Windrush Generation' in the United Kingdom. The Jamaican-born Stuart Hall rightly pointed to the 'outside history that is the inside history of the English' made up global circulation of goods, capital and people (1991: 48). Likewise, in the French context, scholars such as Alain Mabanckou and Achille Mbembe are calling for a 'way out of the Francophone ethnocentrism' (Meddi 2018: 4). Multi-locality is a way of re-mapping new geopolitics and of the exploration of a new dynamic horizon and new prospects of rethinking many key notions such as citizenship, development and sense of belonging. Every point in this new mapping becomes equidistant. There is no specific centre. Globalectics derives from an equidistant positioning whereby a point can serve as centre or periphery in a world of artificial boundaries.

My hope is that the book has made the reader consider not only the academic relevance of memory but also the relevance of exploring new forms of social and political engagement through remembrance narratives. Certainly, not everyone takes part in local associations, nor do they remain actively engaged with the ancestral land. Nevertheless, crashing the divide between the space of Francophonie and the Commonwealth landscape, Eurocentric French and British hegemonies and communities abroad of Caribbean origin, gives the academic relevance that memory deserves as a potent catalyst into the field of critical geographical and cultural studies.

Bibliography

This bibliography is an effort to list main publications about Caribbean migration and, in particular, those related to Haiti and Jamaica. It is based on eight broad themes: general references (history, geography, political economy, geopolitics); methodology; international migrations; transnationalism, diaspora politics; citizenship and memory; migrants' associative practices; networks, development; Haitian/Jamaican migrations; and the Global South.

Achille, T. E. (2007). *Les Haïtiens et la double nationalité*. Montréal: Editions du Marais.

Adélaïde-Merlande, J. (1991). *La Guadeloupe. Les Antilles et la Révolution française. Itinéraires. Avant-propos M. Bangou*. Guadeloupe: Office Régional du Patrimoine Guadeloupéen.

______. (2002). *Histoire contemporaine de la Caraïbe et des Guyanes. De 1945 à nos jours*. Paris: Karthala.

Aggoun, A. (2009). *Enquêter auprès des migrants Le chercheur et son terrain*. Paris: L'Harmattan.

Al-Ali, N., and Khalid, K. (eds) (2002). *New approaches to migration? Transnational Communities & the transformation of home*. London: Routledge.

Alaux, J.-P. (1992). 'La France coupe la route de l'exil', *Plein Droit, GISTI*, 18–19.

Aldred, J. (1998). *The Jamaican Diaspora A People of Paine & Purpose*. Kingston: Morgan Ministries International, Inc.

______. (2012). 'The importance of Faith in the Jamaican community in the UK', *Jamaica Times* (June), 17.

Alexandre, G. (2001). 'La question migratoire entre la Répu:blique dominicaine et Haïti: matériaux pour une proposition de politique, Santo Domingo', *OIM*.

Alexis, S. (1998). 'Typologie historique de la migration haïtienne en France (1ère partie)', *Diaforama* 5 (Sept.), 30–31.

______. (1999a). 'Communauté Haïtienne en Île de France: Une approche anthropologique Nouvelle', *Diaforama* 6 (June), 30–32.

______. (1999b). 'Typologie historique de la migration haïtienne en France (2ème partie)', *Diaforama* 6 (Feb.), 48–49.

Allen, R. M. (2009). 'Anthropology in an era of Transformation in the Netherlands Antilles, *NAAM Report*, pp. 11–17, Executive councils of the Island Territories, Central Government of the Netherlands Antilles, Curacao.

Allwood, G., and Wadia, K. (2011). *Refugee women in Britain and France*. Manchester: Manchester University Press.

Alscher, S. (2010). 'Île d'Hispaniola Des voisins inégaux confrontés à des défis semblables: dégradation environnementale et migration', *Hommes & migrations*, 1284 (3), 84–95.

Amoranitis, S., and Manço, A. (eds) (2010). 'Migration et développement en Europe: politiques, pratiques et acteurs', *EUNOMAD*, 130.

Andrews, M. (2013). *Doing Nothing is not an Option. The radical lives of Eric and Jessica Huntley*. London: Krik Krak.

Anglade, G. (1969). *Contribution à l'étude de la population d'Haïti. L'évolution démographique. La répartition géographique*, Doctoral dissertation, Université de Strasbourg.

______. (1975). *L'espace haïtien*, Montréal: Presses Université du Québec.

______. (1979). 'Haiti: A Space to be invented', *Caribbean Alternatives*, 55–61.

______. (1982a). *Atlas Critique d'Haïti*. Montréal, Groupe d'Études et de Recherches Critiques d'Espace et CRC de l'Université de Montréal.

______. (1982b). *Espace et liberté en Haïti*. Groupe d'Études et de Recherches Critiques d'Espace et CRC de l'Université de Montréal.

______. (1983). *Eloge de la Pauvreté*, Montréal: Les éditions ERCE.

______. (2001). 'Deux concepts en bout de piste. Dixième département et double nationalité', pp. 18–20. *Le Nouvelliste*, Cahier spécial: Bilan Économique.

______. (2004). *L'espace d'une génération, entretiens avec Joseph Josy Lévy*. Montréal: Liber.

______. (2007). 'L'autre Haïti possible', L'Hebdo de G. Anglade, *Le Nouvelliste*, 103.

Anglade, M. (1986). *L'autre moitié du développement à propos du travail des femmes en Haïti*, Paris: Editions des Alizés.

Anteby-Yemini, L., and Berthomière, W. (2005). 'Avant-propos – Di[a]spositif: Décrire et comprendre les diasporas', in L. Antenby-Yemini, W. Berthomière and G. Sheffer (eds), *Les diasporas. 2000 ans d'histoire*, Rennes: PUR, 9–19.

Antonin, A. (1995). *Simon Bolivar et ses idéaux*. Editions Centre Pétion Bolivar.

Appadurai, A. (2001). *Après le colonialisme: les conséquences culturelles de la globalisation*, Paris: Edition Payot.

Archer, C. (2000). *The Forging of a Caribbean Ethnic Identity in New York City Politics*, PhD Dissertation, City University of New York.

Arthuis, W., and Toussaint, Emmanuel (2008). *Radiographie de la communauté protestante haïtienne de France*. Paris, Éditions de l'Alliance.

Aubourg, C. D. (2002). *Des Haïtiens à Paris,* Mémoire de maîtrise d'ethnologie, Université René Descartes, Paris V.

______. (2003). *Oralité et expression scénique dans les lieux de rassemblement des Haïtiens à Paris,* Mémoire de DEA, Anthropologie, Paris, EHESS.

Audebert, C. (2002). 'Le fait religieux dans l'insertion et l'organisation spatiale de la communauté haïtienne de Miami', *Géographie et Cultures,* 43, L'Harmattan, 107–127.

______. (2003). *Les Haïtiens à Miami: l'insertion socio-spatiale d'une population antillaise dans la ville étatsunienne,* PhD Thesis (Geography) Schœlcher, UAG, 2 vols.

______.(2004). 'La constitution de l'espace migratoire international des Jamaïcains: de la dépendance coloniale à la déterritorialisation', in *Haïti dans le Monde, Cahiers des Anneaux de la Mémoire,* 7, 70–84.

______. (2012). *La diaspora haïtienne Territoires migratoires et réseaux transnationaux,* Paris: PUR.

Audebert, C., and Desse, M. (2009). Migrations, mobilités et constructions identitaires caribéennes. *Etudes caribéennes,* 8, 197.

Audebert, C., and Domenach, H. (2008). 'Les migrants caribéens: réseaux et descendance', *REMI,* 24(1) 169.

Audebert, C., and Ma Mung, E. (eds) (2007). *Les migrations internationales: enjeux contemporains et questions nouvelles,* Bilbao, Université Deusto.Augeron, M., et al. (2012). *Floride, un rêve français (1562–1565).* Trouville-sur-mer: Librairie des Musées.

Badie, B. (2014). *Le temps des Humiliés. Pathologie des relations internationales.* Paris: Odile Jacob, 246 p

Badie, B., and Dewitte, P. (1997). 'Quelle citoyenneté à l'heure de la mondialisation?', *Hommes et Migrations,* 1206, 5–13.

Badie, B., and Wihtol de Wenden, C. (1994). *Le défi migratoire Questions de relations internationales.* Paris: Presses fondation nationale.

Badie, B., et al. (2008). *Pour un nouveau regard sur les migrations. Construire une gouvernance mondiale.* Paris: La Découverte.

Balch, E. G. (1927). *Occupied Haiti.* New York: Writers Publishing Company.

Balibar, E. (2001). 'Vers la citoyenneté imparfaite', in E. Balibar (ed.), *Nous, citoyens d'Europe? Les frontières, l'Etat, le peuple,* pp. 206–215. Paris: La Découverte.

Bancel, N., and Blanchard, P. (2000). 'Le colonialisme, "un anneau dans le nez de la République", L'héritage colonial un trou de mémoire', *Hommes & Migrations,* n° 1228 (Nov./Déc.), 80–92.

Bashi, V. F. (2007). *Survival of the Knitted. Immigrant Social Networks in a Stratified World,* Stanford, CA: Stanford University Press.

Bastide, R., Morin, F., and Raveau, F. (1974). *Les Haïtiens en France.* Paris: Mouton.

Bastien, A. (1985). 'La présence haïtienne à Montréal. Évolution démographique et Spatiale', *Cahiers québécois de démographie*, 14 (2), 241–257.

______.(1986). *Distribution spatiale et structure démographique: la Population haïtienne dans la région montréalaise (1971–1982)*, mémoire, Université de Montréal.

Bauböck, R. (1994). *Transnational citizenship*. London: Edward Elgar Publishing.

Bauböck, R. et al. (eds) (2006). *Acquisition and Loss of Nationality, Volume 1: Comparative Analyses*. Amsterdam: Amsterdam University Press.

Bazile, P. (1998–2003). *L'annuaire de la communauté haïtienne France*. Montreuil: Rotographie.

Bazin, M. L. (2005). 'Une politique de la migration', *Le Nouvelliste*, No. 37763: 39.

______. (2007). 'La double nationalité', *Le Nouvelliste*, No. 37765: 33–35.

Beau, S., and Weber, F. (1997). *Guide de l'enquête de terrain*. Paris: La Découverte.

Beauchemin, C., Hamel, C., and Simon, P. (2010). *Trajectoires et Origines. Enquête sur la diversité des populations en France*, INED/ INSEE, documents n°168.

Beaurepaire, P.-Y., and Marzagalli, S. (2010). *Atlas de la Révolution française. Circulation des hommes et des idées 1770–1804*. Paris: Editions Autrement (coll. Atlas/Monde).

Beauvoir-Dominique, Rachel (1991). *L'ancienne Cathédrale de Port-au-Prince. Perspectives d'un vestige de Carrefours*. Editions Henri Deschamps, ISPAN, CRU-T.

Bebel-Gisler, D., and Hurbon, L. (1975). *Cultures et pouvoir dans la Caraïbe. Langue créole, vaudou, sectes religieuses en Guadeloupe et en Haïti*. Paris: L'Harmattan.

Béchacq, D. (2010). *Pratiques migratoires entre Haïti et la France. Des élites d'hier aux diasporas d'aujourd'hui*. Paris: EHESS, PhD Thesis.

Beckford, G. (1972). *Persistent Poverty: Underdevelopment in Plantation Economies of the Third World*. London: Oxford University Press.

Beckford, G., and Witter, M. (1982). *Small Garden, Bitter Weed: Struggle and Change in Jamaica*. Morant Bay: Maroon Publisher House.

Beckford, R. (1998). *Towards a Dread Pentecostal Theology: The Context of a viable Political Theology within Black Pentecostal Churches in Britain*, PhD Dissertation, University of Birmingham.

______.(2002). *Jesus is Dread: Black Theology and Black Culture in Britain*. London: Longman and Todd Ltd.

Beckles, H. (2015). 'The Economics of Black Freedom, Reparation and Anti-Colonial Development', Sir Arthur Lewis Centenary Lecture Ramphal Institute.

Belaise, M. (2004). 'Le facteur religieux, hypothèse explicative de la dynamique entrepreneuriale de la communauté haïtienne de la Guadeloupe et de la Martinique', *Mofwaz*. Matoury: Ibis Rouge Éditions (6), 53–61.

______. (2010). 'Les Églises protestantes haïtiennes de Martinique: des tiers-Lieux d'éducation et d'intégration ?', *Études caribéennes* (16), <http://etudescaribeennes.revues.org/4600> accessed 10 June 2010.

Bellegarde, D., Price-Mars, J., and Laleau, L. (1935). 'Haïti ou Hispaniola', *Revue de la Société d'Histoire et de géographie d'Haïti*, 6 (17), 1–7.

______. (1947). *Dessalines a parlé*. Société d'Editions et de librairie, Haïti.

Bellegarde-Smith, P. (1980). 'Haitian Social Thought in the Nineteenth Century Class Formation and Westernization', *Caribbean Studies*, 20 (1), 5–33.

______. (1990). *Haiti: the Breached Citadel*. Boulder, CO & London: Westview.

Belot, M. (2011). 'L'intégration haïtienne en Guadeloupe', L-A. Joint and J. Mérion (eds), *L'immigration haïtienne dans la Caraïbe*, pp. 97–110. Guadeloupe: Éditions Nestor.

Benson, L. (2014). 'A Queen in Diaspora: The Sorrowful Exile of Queen Marie-Louise Christophe', *The Journal of Haitian Studies*, 20 (2), 90–125.

Bernadel, P-L. (2010). *Itinéraire d'un enfant de paysan ayisyen. De Fonds-des-Blancs À Menilmontant*. Paris: RIP éditions.

Bernard, J., Jr (2010). *Histoire des colonies arabe et juive d'Haïti*. Port-au-Prince: Editions Henri Deschamps.

______.(2011a). *Histoire de la Colonie Allemande d'Haïti*. Port-au-Prince: Editions Henri Deschamps.

______. (2011b). *Histoire de la Colonie Italienne d'Haïti*. Port-au-Prince: Editions Henri Deschamps.

______. (2013). *Histoire Juive d'Haïti*. Port-au-Prince: Editions Henri Deschamps.

Bernèche, F. (1983). 'Immigration et espace urbain. Les regroupements de population haïtienne dans la région métropolitaine de Montréal', *Cahiers québécois de démographie*, Montréal, 12 (2), 295–322.

Berthomière, W., and Hily, M.-A. (2006). 'Décrire les migrations internationales. Les expériences de la co-présence', *REMI*, 22 (2), 67–82.

Bertini, B., et al. (2010). *Une citoyenneté aux multiples visages territoriaux (CMVT)*, Rapport final, Migrinter/GRDR, PUCA.

Bertossi, C. (2002). 'Dilemme de la citoyenneté Intégration/anti-discrimination en Europe, le cas français'. Liège: *Colloque CEDEM*.

______. (2007). 'French and British models of integration: public philosophies, Policies & State institutions', ESCR Centre on Migration, Policy & Society, 46. Oxford, 7–9.

______. (2008). 'The Regulation of Migration: A Global Challenge', *Politique étrangère*, Hors-Série, special issue, pp. 189–202.

______. (2016). *La citoyenneté à la Française*. Paris: CNRS Editions.

Bertossi, C., and Wihtol de Wenden, C. (2007). *Les Couleurs du drapeau: l'armée française face aux discriminations*. Paris: Robert Laffont.

Besson, J. (2016). *Transformations of Freedom in the Land of the Maroons*. Ian Randle Publishers.

Bilby, K., and N'Diaye, D. B. (1992). *Creativity and Resistance: Maroon Communities in the Americas*. Washington, DC: Smithsonian Institution.

Black, C. V. (1961). *History of Jamaica*. London: Collins Clear Press.

Blanchet, A., et al. (1992). *L'entretien dans les sciences sociales*. Paris: Nathan.

Blancpain, F. (2001). *Un siècle de relations financières entre Haïti et la France (1825–1922)*. Paris: L'Harmattan.

Blaxter, L., Hughes, C., and Tight, M. (2006). *How to Research* (3rd edn), Berkshire, UK: Maidenhead.

Bloncourt, G., and Löwy, M. (2007). *Messagers de la Tempête. André Breton et la Révolution de Janvier 1946 en Haïti*. Paris: Le Temps des Cerises.

Blum, W. (ed.) (2004). 'La Jamaïque, 1976–1980', in *Les Guerres scélérates*, Paris: L'Aventurine/Parangon, 3, 280–284.

Boisseron, M. (2007). 'Émigration haïtienne en République Dominicaine: source de profits, source de conflits', *Dynamiques migratoires de la Caraïbe*, 6, Géode Caraïbe. pp. 21–34. Karthala.

Bonacci, G. (2009). 'A contre-courant: les mobilités caribéennes vers l'Ethiopie', in C. Audebert and M. Desse (eds), *Etudes Caribéennes, Migrations, mobilités et constructions identitaires caribéennes*, pp. 39–48. Paris: éditions Publibook.

Bonilla, Y. (2012). 'Le syndicalisme comme marronnage: épistémologies du travail et de l'histoire en Guadeloupe', in J. C. William et al. (eds), *Mobilisations sociales aux Antilles*, pp. 77–94. Paris: Édition Karthala.

Bonnin, J. C. (2005). *Les Templiers de La Rochelle*. A & T Editions.

Borda, P., Maurn, A., and Manioc, O. (2008). 'L'immigration intra-caribéenne. Des faits aux enjeux économiques', *Hommes et migrations*, 1274, 10–20.

Bordes-Benayoun, C. (2012). 'La diaspora ou l'ethnique en mouvement', *REMI*, 28 (1), 13–31.

Bordes-Benayoun, C., and Schnapper, D. (2006). *Diaspora et nations*. Paris: Odile Jacob.Boucard, A. (2006). *La communauté haïtienne de Montréal, essai de sociologie*, Fondation littéraire Fleur de Lys. Québec: Laval.

Bougerol, C. (2010). 'Une rumeur à la Guadeloupe. De certaines pratiques supposées des Haïtiens', *Terrain*, 54, 130–139.

Brady, P. (2010). 'It's a love Haiti relationship!'. Birmingham, *West Midlands News, The Voice*, 2–8 December, p. 16.

Brathwaite, K. (2005). *The Development of Creole Society in Jamaica 1770–1820* (3rd edn). Kingston: Ian Randle Publishers.

Braud, P. (2004). *Penser l'Etat*. Paris: Editions du Seuil.

Bret, B. (2001). *Le Tiers-Monde Croissance, développement, Inégalités*, 3rd edn. Paris: Ellipses.

Brettell, C. B., and Hollifield, J. F. (eds) (2008). *Migration Theory: Talking across Disciplines*, New York & London: Routledge.

Bridge, C., and Fedorowich, K. (2003). *The British World: Diaspora, Culture, and Identity*. London: Routledge.

Brière, J.-F. (2008). *Haïti et la France 1804–1848, Le rêve brisé*. Paris: Karthala.

Brierre, J. (1953). *Dessalines nous parle*, Port-au-Prince: Deschamps.Brisson-Lamaute, N.(2000). *L'économie informelle en Haïti: le cas de l'aire métropolitaine de Port-au-Prince*, PhD Thesis, Paris X.

Brodwin, P. (2003). 'Pentecostalism in Translation: Religion & the Production of Community in the Haitian Diaspora', *American Ethnologist* 30 (1), 85–101.

Browder, A. (2006). *Egypt on The Potomac*. Washington, DC: IKG.

Brubaker, R. (1992). *Citizenship and Nationhood in France and Germany*. Cambridge, MA: Harvard University Press.

Brudey, S. (2008). *Les relations entre Cuba et la Guadeloupe (département français d'Amérique 1960–2008)*, Master's Thesis, IHEAL, Sorbonne-Nouvelle, Paris.

Bulmer-Thomas, V. (2012). *The Economic History of the Caribbean since the Napoleonic Wars*. Cambridge: Cambridge University Press.

Burac, M. (1994). *Guadeloupe, Martinique et Guyane dans le monde américain*. Paris: Édition Karthala.

Burnard,T. (2008). 'The Atlantic Slave Trade and African Ethnicities in Seventeenth Century Jamaica', in Richardson et al. (eds), *Liverpool and Transatlantic Slavery*. Liverpool: Liverpool University Press.

Burton, R., and Reno, F. (1995). *French & West Indian*. London: Macmillan.

Butel, P. (1992). *Les Négociants bordelais, l'Europe et les îles au XVIIIe siècle*, Aubier-Montaigne, collection historique, 431p.

Byron, M. (1998). 'Migration, Work and Gender: the case of post-war labour migration from the Caribbean to Britain', in M. Chamberlain (ed.), *Caribbean Migration: Global Identies*, pp. 217–231. London: Routledge.

Byron, M., and Condon, S. (1996). 'A comparative study of Caribbean return migration from Britain and France: towards a context dependent explanation', *Transactions of the Institute of British Geographers*, 91–104.

______. (2008). *Migration in comparative perspective Caribbean communities in Britain and France*. London: Routledge.

Cabanes, A., and Robbes, A. S. (2003). La coopération décentralisée: comment s'extraire des contingences de l'intérêt public local? *Actualité juridique – droit administratif*, 12, 59–602.

Cadet, C. (1982). *La spécificité des voies de l'accumulation du capital dans l'économie haïtienne*, PhD Thesis, EHESS, Paris.

______. (2008). *La politique de subvention de l'Etat haïtien face à la crise socioéconomique actuelle*, Policy paper, Port-au-Prince: CAPESC.

______. (2012). 'L'enjeu des petites et moyennes entreprises pour le relèvement de l'économie haïtienne', *Rencontres*, 24–25 CRESFED, pp. 52–55.

Calmont, A. (1993). 'Les Haïtiens en Guyane: une communauté en voie d'intégration ?, *Espaces-Populations-Sociétés*, Université Lille, 427–434.

______. (2007). 'Trajets socio-identitaires chez les jeunes issus de la migration haïtienne en Guyane', *Dynamiques migratoires de la Caraïbe*, Terres d'Amérique, n°6, Paris, Karthala/Géode Caraïbe, pp. 199–220.

______. (2009). 'Le commerce informel international en Haïti: les Madan Sara', *Les littoraux de la Caraïbe*, coll. Terres d'Amérique, Karthala/Géode Caraïbe, pp. 179–194.

______. (2011). 'Histoire de vie d'un couple d'Haïtiens installés en Martinique', L-A Joint and J. Mérion (eds), *L'immigration haïtienne dans la Caraïbe*, Gourbeyre (Guadeloupe): Éditions Nestor, pp. 309–310.

Calmont, A., and Audebert, C. (2007). 'Immigration antillaise et diversité sociale de la population noire aux Etats-Unis', *Dynamiques migratoires de la Caraïbe* (coll. Terres d'Amérique), Karthala/Géode Caraïbe, 77–92.

Calmont, R. (1988). 'L'impact de l'immigration haïtienne en Guyane', *Équinoxe*, Cayenne, 26, 1–48.

Calmont, R., and Gorgeon, C. (1987). 'L'immigration haïtienne en Guyane', *Equinoxe*, Cayenne, 23, 1–16.

Campbell, H. (2007). *Rasta and Resistance*. Hertford, UK: Hansib.

Capone, S. (2010). 'Religions en migration: de l'étude des migrations internationales à l'approche transnationale', *Autrepart*, 54, 235–259.

Casey, M. (2012). *From Haiti to Cuba and Back, Haitians' Experiences of Migration, Labor, and Return,1900–1940*, PhD, University of Pittsburgh.

Casimir, J. (1991). *La Caraïbe Une et Divisible*. Port-au-Prince: coédition Édition Deschamps-CEPALC.

______. (2004). *Pa bliyé 1804 (Souviens-toi de 1804)*. Port-au-Prince: Lakay.

Casséus, Ernst Pedro. (2004). *Dictionnaire géographique et toponymique de la république d'Haïti*. Port-au-Prince: Deschamps. Contribution to the 2nd Edition (2016). Preface Clara Rachel Eybalin Casséus.

Casséus, J. V. (2003). *Karikati*, political cartoons (J. V. KCUSH) in journal *Libète*, Port-au-Prince: Edition Deschamps.

______. (2006). Cartes administratives d'Haïti, Bureau du IVème recensement général de la population et de l'habitat, division cartographie, Haïti.

Casséus, R. M. (1985). *Approche de gestion en secteur informel: le cas du marché Tête Bœuf*, Thesis, INAGHEI, Port-au-Prince.

______. (1987). *Stratégie alternative de développement, secteur informel et l'Etat dans une petite économie latino-américaine: le cas d'Haïti*, Master's Thesis, Centre de recherche et d'enseignement économiques (CIDE), Mexique.

Casséus, R. M., and Mathon, D. (2004). 'Les finances publiques dans la Décentralisation', 25–36. Port-au-Prince: Publication INESA-Le Nouvelliste.

Castera, G. (2005). *Un pays sur parole*, Riveneuve Continents 2 (Printemps), pp. 121–124.

Castles, S. (2004). 'Why Migration Policies Fail', *Ethnic and Racial Studies*, 27 (2), 205–227.

Castor, S. (1988a). 'Le massacre de 1937 et les relations haïtiano-dominicaines'. *Migrations et relations internationales: le cas haïtiano-dominicain*. Port-au-Prince: CRESFED.

______. (1988b). *L'occupation américaine d'Haïti* (1er prix 1987 de la société haïtienne d'histoire). Port-au-Prince: Deschamps.

______. (2004). 'La question migratoire dans la Caraïbe à l'aube du 21ème siècle, *Alternatives Sud*, vol. 11, pp. 153–164.

Castor, S., et al. (2012). 'Le défi de la reconstruction', revue *Rencontre*, CRESFED, 24–25, 5–17.

Catanese, A. V. (1999). *Haitians: migration and diaspora*. New York and Boulder, CO: Westview Press.

Cauna, J. de (1985). 'Une famille transatlantique: les Fleuriau', *Les Cahiers de Framespa* <https://journals.openedition.org/framespa/1152?lang=en>.

______. (1994). 'La diaspora des Colons de Saint-Domingue et le monde créole:le cas de la Jamaïque', *Revue Française d'Histoire d'Outre-Mer*, 333–359.

______. (2012). 'Une émigration intra-caribéenne post-révolutionnaire méconnue: la diaspora des réfugiés de Saint-Domingue (fin XVIIIe – début XIXe siècle)', in M. Dalmace (ed.), *La mer Caraïbe, espace de migrations*, pp. 209–232. Bordeaux: Presses Universitaires de Bordeaux.

Cazenave, J. (2006). *L'analyse du fardeau de la dette extérieure publique selon la Banque Mondiale*, Working paper (économie et gestion).

Célimène, F., and Legris, A. (eds) (2002). *L'économie de l'esclavage colonial*. Paris: CNRS Éditions (coll. Biblis).

Césaire, A. (1958). *Et les chiens se taisaient*, Paris: Présence Africaine.

Ceyrat, A. (2009). *Jamaïque La construction de l'identité noire depuis l'indépendance*. Paris: L'Harmattan.

Chalifoux, J. J. (1988a). *Chercher 'Lavi' en Guyane française. Témoignages d'Haïtiens et d'Haïtiennes immigrés*, Département d'Anthropologie. Québec: Université de Laval.

Chamberlain, M. (1998). *Caribbean Migration: Globalised Identities*, London: Routledge.

Chancy, M. (2010). 'L'aide pourrait empoisonner la relance d'Haïti', journal canadien le *Soleil*, 2 February, dossier special séisme Haiti.

Charef, M., and Gonin, P. (eds) (2005). *Emigrés/Immigrés dans le développement local.* Agadir: Editions Sud Contact.

Charier, A. (2002). 'L'immigration haïtienne au Venezuela: identité et Intégration', *Pouvoirs dans la Caraïbe*, Schœlcher,11, p. 175–197.

Charles, C. (1992). 'Transnationalism in the Construct of Haitian Migrants' Racial Categories of Identity in New York City', in N. Glich Schiller et al. (eds), *Towards a Transnational Perspective on Migration*. New York: The New York Academy of Sciences.

Charles, C., and Exulien, J.-C. (2003). 'Pour Honorer les Haïtiens tombés à Savannah en 1779', *Haïtiens Aujourd'hui*, pp. 25–27.

Charles, J. (2006). 'Bahamas Haitians Face Identity Crisis'. *Miami Herald*, 18 April, A1.

Chatelain, M. (2015). *South Side Girls. Growing up in the Great Migration*. Durham, NC: Duke University Press.

Chauvet, M. (2012). 'Où en est Haïti un après le séisme ?', in *Haïti, Réinventer l'avenir*, J.-D. Rainhorn (ed.), pp. 125–132. Paris: Editions MSH/Université d'Etat d'Haïti.

Chérubini, B. (2000). 'Sectes et Églises face aux immigrés haïtiens en Guyane française', in L. Hurbon (ed.), *Le phénomène religieux dans la Caraïbe*, pp 207–218. Paris: Karthala.

Chevannes, B. (2001). 'The Affirmation of Jamaican Diasporic identity: the Metaphor of Yaad', in P. Taylor (ed.), *Nation Dance: Religion, Identity, and Cultural Difference in the Caribbean*. Bloomington: Indiana University Press.

Chevannes, B., and Ricketts, H. (1997). 'Return Migration and Small Business Development in Jamaica', in Patricia Pressar (ed.), *Caribbean Circuits: New Directions in the Study of Caribbean Migration*, pp. 161–175. New York: CMS.

Chicot, P.-Y. (2008). 'L'immigration clandestine en Guyane française. Entre contestation et nécessité d'adaptation du droit'. *Hommes et migrations*, 1274, 76–89.

Chivallon, C. (1995). Les espaces de la diaspora antillaise au Royaume-Uni. Limites des concepts socio-anthropologiques, *Les annales de la recherche urbaine*, pp. 198–209.

______. (2002). 'Présence antillaise au Royaume-Uni, Diasporas caribéennes', *Hommes & Migrations*, n°1237 (mai-juin), pp. 62–69.

Chomsky, N., and Farmer, P. (2004). *Getting Haïti right this time*, Common Courage Press, 180.

Clarke, G., and Wallsten, S. (2004). 'Do Remittances Protect Household in Developing Countries against Shocks? Evidence from a Natural Disaster in Jamaica', *Mimeo*, World Bank.

Clerismé, R. (2003). *Main-d'œuvre haïtienne, capital dominicain. Essai d'anthropologie Historique*. Paris: L'Harmattan.

Cohen, J. (2002a). 'Diaspora et incorporation: présences publiques des Caribéens aux Etats-Unis', *Diasporas caribéennes, Hommes et migration*, n° 1237, 82–90.

______. (2002b). 'Entre Caraïbes et métropoles: parcours diasporiques et Citoyennetés', *Diasporas caribéennes, Hommes et migration*, 1237.

Cohen, R. (1993). 'La diaspora d'une diaspora: le cas des Antilles', in R. Gallissot (ed.), *Pluralisme culturel en Europe. Culture(s) européenne(s) des diasporas,* p. 61–77. Paris: L'Harmattan.

______. (1998). 'Cultural Diaspora: the Caribbean case', in Chamberlain, M. (ed.), *Caribbean Migration: Globalised identities,* 21–35. London/New York: Routledge.

______. (2008). *Global Diaspora: An Introduction*, 2nd edn. London, Routledge.

Colimon, S. (1999). Diaspora, CD *Le mensonge*, Québec: production Scoda.

Collier, P. (2008). *The Bottom Billion Why the Poorest Countries Are Failing and What can be Done About It.* Oxford: Oxford University Press.

______. (2009). *Haïti: des catastrophes naturelles à la sécurité économique, Rapport au Secrétaire Général de l'Organisation des Nations Unies.*

Collyer, M. (2014). 'A geography of extra-territorial citizenship: Explanations of external voting', *Migration Studies*, 2 (1), 55–72.

Coly, A. (2017). 'Can globalization un-doom the category of Francophone Literatures? Lessons from Alain Mabanckou', keynote presentation, Annual Conference of Society for Francophone Postcolonial Studies, Senate House, UK.

Comissiong, D. A. (2013). *It's the Healing of the Nation. The Case for Reparations in an Era of Recession & Re-colonisation*, Barbados: Caribbean Chapters Publishing.

Condon, S. (1994). 'L'accès au logement: filières et blocages. Le cas des Antillais en France et Grande-Bretagne', *Population*, 49 (2), 522–530.

Conerly, Jennifer Yvonne (2013). *Your Majesty's Friend': Foreign Alliances in the Reign of Henri Christophe*, Master's Thesis (History), New Orleans, LA: University of New Orleans.

Confiant, R. (2009). 'Brèves considérations sur l'indigénisme ...', in Jean Price-Mars [présentation de R. St-Eloi], *Ainsi Parla L'Oncle suivi de Revisiter L'Oncle*, pp. 323–327. Montréal, Québec: Mémoire d'Encrier.

Constant, F. (1987). 'La politique française de l'immigration antillaise (1946–1987)', *REMI*, 3 (3), 9–30.

______. (1991). 'Les nationaux de la France d'Outre-mer: des citoyens comme les autres?' in D. Colas et al., *Nationalité et Citoyenneté.Perspectives en France et au Québec, pp.* 243–261. Paris: PUF.

______. (1998a). 'French Republicanism under challenge: white minority béké power in Martinique and Guadeloupe' in H. Johnson Karl Watson (ed.), *The White Minority in the Caribbean,* Princeton, NJ: Marcus Wiener Publishers, Chapter 9.

______. (1998b). *La Citoyenneté.* Paris: Edition Montchrestien.

Converse, D. (2012). *Post-earthquake Challenges and Successes of Haitian Immigrants Residing in Brooklyn, NY*, Paper, HSA, CUNY.

Conway, D. (1992). 'Migration trends in the Americas', in Tom L. Martinson & Susan Booker-Gross (eds), *Revisiting the Americas: Teaching and Learning the Geography of the Western Hemisphere*, pp. 148–151.

Cooper, C. (1995). *Noises in the Blood*, Durham, NC: Duke University Press.

Cooper, D. W. (1985). 'Migration from Jamaica in the 1970s: Political Protest or Economic Pull?' *International Migration Review*, 19 (4), 728–745.

______. (2008). *Global Diaspora: An Introduction*, 2nd edn. London, Routledge.

Coquery-Vidrovitch, C. (1999). 'Lieux de mémoire et occidentalisation', in J.-P. Chrétien, J.-L. Triaud, Karthala (eds), *Histoire d'Afrique: Les enjeux de mémoire*, pp. 378–379.

Corsi, J. (2016). *Partners in Crime*. Washington, DC: WND Books.

Corten, A. (1992). 'Les réfugiés haïtiens à l'assaut de l'Amérique', *Le Monde diplomatique* (février).

Cottias, M. (1998). 'Introduction', in *Anthologie*, pp. 3–15. Marseille: Angone.

Cottret, B. (2007). *Histoire de l'Angleterre*. Paris: Editions Tallandier.

Courtin, C. (2007). 'Le Codéveloppement: un alibi pour des politiques restrictives', *La Revue internationale et stratégique* (hiver), 68, 42–48.

Craton, M. (1995). 'The Bahamian Self and the Haitian Other: The Migration of Haitians to and through the Bahamas, 1950–2000', *Immigrant and Minorities* 14 (3), 265–288.

Crusol, J. (2007). 'Abolition et transition post-esclavagiste dans les sociétés insulaires de la Caraïbe', in Éric Lambourdière (ed.), *Les Caraïbes dans la géopolitique mondiale*, pp. 49–86. Paris: Éllipses Editions (coll. Carrefours).

Cunningham, R. (2006). 'Impacting Peace Building and Development in Jamaica: Addressing Challenges and Opportunities Encountered by the Jamaican Diaspora', Conference, Toronto.

Curtin, P. D. (1955). *Two Jamaicas The Role of Ideas in a Tropical Colony 1830–1865*, Wesport, CT: Greenwood Press.

D'Almeida-Topor, H. (2016). *Les Amazones*. Besançon: la Lanterne magique.

D'Ans, A. M. (2009). 'Détérioration écologique du milieu dominguois', La Problématique de l'environnement en Haïti aujourd'hui, *Conjonction*, 221/2.

Daguerre, A. (2009). 'L'intégration des migrants au Royaume-Uni', in C. Jaffrelot et C. Lequesne (eds), *L'enjeu mondial*, pp. 207–215. Paris: Presses de Sciences Po.

Dalmace, M. (ed.) (2012). *La mer Caraïbe, espace de migrations*. Presses Universitaires de Bordeaux-Pessac.

Dalmas, A., Wante, C. E. P (1814). *Histoire de la révolution de Saint-Domingue: depuis le Commencement des troubles, jusqu'à la prise de Jérémie et du Môle St Nicolas. D'un mémoire sur le rétablissement de cette colonie*. Paris: imprimerie de Mame Frères.

Dalphinis, M. (1993). 'The creole language towards the twenty-first century'. Feature address for the Creole Festival (26 October), Castries, St Lucia.

Daniel, J., and Arnaud, L. (eds) (2005). 'L'UE et la construction des identités: l'exemple des populations caribéennes', in *Les minorités ethniques dans l'Union européenne: politiques, mobilisations, identités*, pp. 113–135.

Danroc, G. (1992). 'États-Unis: le retour des refoulés', *Plein Droit*, GISTI 18/9.

Dartigue, E. (1992). *Un Haïtien exceptionnel Maurice Dartigue.* Paris: éditions J'étais Une fois.

Daum, C. (1998). *Les associations de Maliens en France: Migration, Développement et citoyenneté.* Paris: Karthala.

Davidson, B. (1969). '"No Place like Home", A study of Jamaica Returning to Kingston, Jamaica', *Race*, 9 (4), 499–509.

De Haas, H. (2006). 'Engaging Diasporas: How Governments and Development Agencies Can Support Diaspora Involvement in the Development of their Origin Countries', *International Migration Institute (IMI)*, Oxford.

______. (2010). Migration and Development: A Theoretical Perspective. *IMR* 44, 1 (Spring), 227–264.

De Ronceray, H. (1979). *Sociologie du fait haïtien*. Québec: Presses de l'université du Québec.

De Tapia, S. (1996). 'Echanges, transports et communications: circulation et champs migratoires turcs', *REMI*, 12, 45–71.

De Wind, J., and Kinley, D. (1988). *Aiding Migration. The Impact of International Development Assistance on Haiti.* Boulder, CO: Westview Press.

Déjean, P. (1978). *Les Haïtiens au Québec.* Montréal, Presses de l'Université du Québec.

______. (1987). *D'Haïti au Québec*, Montréal, Centre international de documentation et d'information haïtienne, caraïbéenne et afro-canadienne (CIDIHCA).

Delachet-Guillon, C. (1996). *La communauté haïtienne en Île-de-France.* Paris: L'Harmattan.

Delafosse, M. (1998). La Rochelle, l'Afrique et l'Amérique, in *L'esclavage dans les collections du Musée du Nouveau Monde*, Imprimerie Aprim, pp. 5–8.

Delancy, K. (2017). 'History to Heritage: A Heritage Assessment of Tarpum Bay, Eleuthera, The Bahamas', paper at *Memory, Migration and Decolonisation in the Caribbean and Beyond, 1804 to the Present*, Senate House, University of London.

Dépestre, M. (2004). 'The Impact of Migration Between Haiti and the English-Speaking Caribbean on Culture & Religion', in *The Haiti-Jamaica Connection*, pp. 55–59. Kingston, Latin American-Caribbean Centre (LACC).

Désaguliers, J.-T. *Un protestant rochelais, fondateur de la Maçonnerie Moderne*, conférence-débat, Bernard Daniel et J.-P. Rulie, Musée du Temple, La Rochelle.

Désir, C. (2006). *Lot bodlo: Across Waters-Haitian Students Search for Identity in U.S. Schools*, PhD dissertation, Harvard University Graduate School of Education.

Desrosiers, T. (1984). Civilisation indienne et Culture haïtienne, Port-au-Prince: Imp. Rodriguez, p. 50.

______. (2001). *Relations Ayitiano-Dominicaines, La vision réaliste de Dr. Jean Price-Mars et du professeur Daniel Fignolé*, Port-au-Prince: Imp. Rodriguez.

Désulmé, M. (2004). 'The impact of Haitian Migrants on Business in Jamaica', *The Haiti-Jamaica Connection*, pp. 44–54. Kingston, LACC Centre.

Deveau, J.-M. (2009). *La Traite rochelaise*. Paris: Editions Karthala.

Demoule, J.-P. 'Les théories des migrations', in *Archéologie des migrations*, sous la direction de Dominique Garcia et Hervé Le Bras, La Découverte Inrap, pp. 25–38.

Dewitte, P. (2002). 'Des citoyens à part entière, ou entièrement à part ?', *Diasporas caribéennes, Hommes et migration*,1237, 1.

Di Méo, G., and Buléon, P. (eds) (2005). *L'espace social. Une lecture géographique des sociétés*. Paris: Armand Colin.

Diène, D. (2001). *From Chains to Bonds: The Slave Trade Revisited*. UNESCO/ Berghahn Books.

Domenach, H. (1986). 'Les migrations intra-caribéennes', *REMI*, 2, 9–24.

______.(1994). 'L'espace migratoire caraïbéen et les départements français d'Amérique', in Maurice Burac (ed.), *Guadeloupe, Martinique et Guyane dans le monde américain*, pp. 175–185. Paris: Éditions Karthala.

______. (1996). 'De la "migratologie"', *REMI*, 12 (2), 73–86.

______. (2002). 'L'évolution au XXe siècle du système démographique et migratoire caraïbéen', *Diasporas caribéennes, Hommes et migration*, 1237, 13.

Dominique, Jan J. (2008). *Mémoire errante*. Québec: Mémoire d'Encrier.

Dominique, Jean Léopold (1986). 'La fin du marronage haïtien: éléments pour une étude des mouvements de contestation populaire en Haïti', *Collectif Paroles*, 32, Montréal (mai-déc.).

______. (1993). 'Qui a peur de la participation', *Chemin Critiques*, 3 (12).

Dominique, M. (1982). 'Les Haïtiens aux Bahamas', *Collectif Paroles*, Montréal, 16, 23 p.

Doraï, K., et al. (1998). *Bilan des travaux sur la circulation migratoire*, E.l. Ma Mung (responsable scientifique), Migrinter.

Dorsainville, M. H. (2006). *Mémoires de la décolonisation*. Montréal: Mémoires d'encrier.

Dorvillier, F. (2007). *Apprentissage organisationnel et dynamique de développement local en Haïti: proposition d'une intelligibilité en termes de production d'un nouvel ordre territorial*, PhD Pop-Développement-Environnement, UCL.

______. (2009). 'La crise haïtienne de développement: entre économie morale et explosion démographique', *Cahiers du CEPODE*, n° 1 (Sept.).

Doucet, R. C. (2012). 'Parc Industriel de Caracol: pour protéger un patrimoinenaturel et culturel en danger', *L'observatoire de la reconstruction*, 2, 8–11.

Douyon, F. (2004). *Haïti de l'Indépendance à la Dépendance*. Paris: L'Harmattan.

Drot, J.-M. (2009). *Le dernier voyage d'André Malraux en Haïti ou la découverte de l'art vaudou*, Musée du Montparnasse et Paradoxe.

Drozdzewski, D.; et al. (2016). *Memory, Place and Identity. Commemoration and remberance of war and conflict*, Abingdon: Routledge.

Dubesset, E. (2012). 'La dynamique géopolitique des migrations caraïbéennes (XVIe-XXe siècles)', in *La mer Caraïbe, espace de migrations*, pp. 349–366. Bordeaux: Presses Universitaires de Bordeaux-Pessac.

Dubois, L. (1998). 'Les réfugiés haïtiens et la politique d'immigration aux Etats-Unis', *Hommes et Migrations*, Paris, 1213.

Dubost, I. (2007). 'Au-delà de l'ethnicité: les chinois à la Martinique', in *Dynamiques migratoires de la Caraïbe*, Terres d'Amérique, 6, 239–252.

Dufoix, S. (2003). *Les diasporas*. Paris: PUF (coll. Que sais-je ?).

Dufoix, S., et al. (2010). *Loin des yeux, près du cœur, Les Etats et leurs Expatriés*. Paris: Presses de Sciences Po.

Dufour, V. (2011). *Haïti: un grand défi pour la coopération internationale et le développement durable*, Centre universitaire de formation en environnement. Quebec: Université de Sherbrooke.

Duhaime, C. (2010). 'La coopération de la Francophonie en faveur d'Haïti: un devoir de fraternité', *Géoéconomie*, 20 (55), 125–130.

Dumas, R. (2008). *An Encounter with Haiti Notes of a Special Adviser*, Port of Spain: Medianet Limited.

Dumont, A. (2007). *La marocanité associative en France. Militantisme et territorialité d'une appartenance exprimée à distance*, PhD Thesis, Geography, Université de Poitiers.

______. (2010). ' L'état des recherches sur les associations de migrants internationaux', *REMI*, 26 (2), 117–137.

Dumont, M. (2011). 'Aux origines d'une géopolitique de l'action spatiale: Michel Foucault dans les géographies françaises', *L'espace Politique*, <http://espacepolitiques.revues.org/index1744.html>.

Dumont, Paul (1973). 'Haïti: de la colonie à la francophonie', *Revue des deux-mondes* (mai), n°5.

Dupuy, A. (1997). *Haiti in the New World Order*. Boulder, CO: Westview Press.

Durand, M.-F. (1994). 'Entre territoires et réseaux', in B. Badie, C. Wihtol de Wenden (eds), *Le Défi migratoire: Questions de relations internationales*, pp. 141–157. Paris.

Duret, P. (1996). 'Développement des services financiers ruraux', in Le Développement rural en Haïti, *Conjonction*, 200, 74–76.

Easterly, W. (2006). *Les pays pauvres sont-ils condamnés à le rester?* Paris: Groupe Eyrolles.

Edmonson, L. (2010). 'The Caribbean Diaspora: Caribbean/ Homeland dialectic and the global Caribbean', in Franklyn, Delano (ed.) *The Jamaican Diaspora*, pp. 344–380. Kingston: Wilson Franklyn and Barnes.

Edwards, J. (1998). *The Jamaican Diaspora A People of Pain & Purpose*. Kingston: Morgan Ministries International, Inc.

Ehrengardt, T. (2009). *Histoire ancienne de l'île de la Jamaïque, de 1494 à 1838*. Editions Natty Dread (coll. Jamaica Insula).

Ekovich, S. R. (2007). 'Les ONG et la politique étrangère des Etats-Unis', *Géostratégiques*, 16, pp. 67–81.

Elden, S., et al. (2015). 'Thinking Past Henri Lefebvre: Introducing "The Theory of Ground Rent and Rural Sociology"', *Antipode*, a radical journal of Geography, vol. 48, 1: 56–66, Wiley, Blackwell.

Elie, J. R. (2006). *Participation Décentralisation Collectivités Territoriales en Haïti*. Port-au-Prince: PAPDA.

______. (2009). *La participation citoyenne en Haïti*, présentation au forum de l'Université du Québec en Outaouais, série; chaire Senghor, n°1 (juin).

Erbès-Seguin, S., and Mooney, M. (2008). 'Structures de médiation et intégration des immigrants haïtiens à Paris', *REMI*, 24 (1), 89–114.

ESOMAR (2017). Conducting Research under the GDPR: Legal Bases, EU guide.

Ethéart, B. (2014). *La problématique foncière en Haïti*. Montreal QC: CIDIHCA.

Etienne, Sauveur Pierre (2011). 'Haïti', *la République dominicaine et Cuba, Etat, économie et société (1492–2009)*, Paris: L'Harmattan, 440 p.

Etzioni, A., and Dubow, F. (1970). *Comparative Perspectives: Theories and Methods*. Boston, MA: Little Brown.

Eybalin Casséus, C. R. (2007). *Analyse du concept de société civile dans la dynamique du conflit haïtien (1986–2006)*, Master's Thesis, Institut Catholique de Paris/AUP.

______. (2008). *Migrants as Transnational Development Actor: The Case of Haitians in Metropolitan France*, Master's Thesis. Public Administration in Strategic Public Policy. The American University of Paris.

______. (2012a). *Diaspora Organizations as Transnational Development Actors: The Case of the Jamaican Diaspora in the United Kingdom*. Paper presented at the *Commonwealth Diaspora Symposium*, University of Liverpool-Commonwealth Foundation, Birmingham City Council.

______. (2012b). *Two years after 12.01.10: Trans-dialogue among two Haitian Communities based in London and Paris*, Paper, 24th Haitian Studies Association, CUNY, New York.

______. (2013). 'Le rôle de la politique transnationale dans le développement du pays d'origine', Paper presented at the *Rethinking Caribbean Transnational Politics: The Challenge of Bringing Development*, 38th CSA, Grenada.

______. (2015). Transnationalisme Associatif Haïtien et Jamaïcain. Géographie du Développement local et politiques diasporiques. Saarbrücken: Editions Universitaires Européennes.

______. (2017a). 'Pushing the Boundaries of Mediterranean Configurations: Refugees as Actors', *Arab council for the Social Sciences Conference*, Beirut. Lebanon.

______. (2017b). 'Decolonial mapping of memory through the path of Pride and Grandeur', Paper, Institute Historical Research (IHR), Senate House, London.

______. (2018a). 'A decolonising Caribbean: Resurrecting the ancestors alive through Tongue', panel, Caribbean Studies Association Conference (CSA), Havana, Cuba.

______. (2018b). 'Framing an Operational South-south Cooperation with China: Post-earthquake Haiti', paper, Caribbean Annual Conference (CSA), Havana, Cuba.

______. (2018c) 'Conflicting Memories, Cross-border displacements, and climate change', panel chair, Society for Caribbean Studies (UK) Conference, London, UK.

Faini, R. (2007). 'Migrations et transferts de fonds. Impact sur les pays d'origine', *Revue d'économie du développement*, 21 (2–3), 153–182.

Faist, T. (1998). 'Transnational social spaces out of international migration: Evolution, significance and future prospects', *Archives Européennes de Sociologie*, 39 (2), 213–247.

______. (2000). *The volume and dynamics of international migration and transnational social spaces*. Oxford: Clarendon Press.

______. (2004). *Transnational social spaces: agents, networks, and institutions*. Burlington: Ashgate.

______. (2012). 'Toward a Transnational Methodology: Methods to Address Methodological Nationalism, Essentialism, & Positionality', *REMI*, 28 (1).

Fardin, D. (2018a). *Jadis et naguère dans la cité. Les balbutiements de la démocratie sous l'ère jeanclaudiste (1971–1986)*, volume I, Port-au-Prince: éditions Fardin.

______. (2018b). *Jadis et naguère dans la cité. La bamboche démocratique à l'enseigne de l'occupation étrangère (2003–2005)*, volume II, Port-au-Prince: éditions Fardin.

Farmer, P. (2003). *The Uses of Haiti*. Monroe, ME: Common Courage Press.

Fass, S. (1988). *Political Economy in Haiti: The drama of survival*, Transactions Books.

Favell, A. (2001). *Philosophies of Integration, Immigration and the Idea of Citizenship in France and Britain*. Basingstoke: Palgrave-Macmillan.

Favre, A. (2006). *Globalisation et métissage. Approche comparée de la population antillaise en France et en Grande Bretagne* Paris: L'Harmattan.

______. (1985). 'Race and Color: Jamaican Migrants in London and New York City', *IMR Review*, 19 (4), 708–727.

Fecteau, J.-M., and Harvey, J. (eds) (2005). *La régulation sociale entre l'acteur et l'institution*. Quebec: Presses de l'Université du Quebec.

Ferguson, N. (2002). *The Cash Nexus: Money and Power in the New World, 1700–2000*. New York: Basic Books.

Fick, C. (1990). *The Making of Haiti: The Saint Domingue Revolution from Below*. Knoxville: University of Tennessee Press.

Fignolé, Jean-Claude. (1985). *Les possédés de la pleine lune*. Paris: Editions Seuil.

______. (2011). 'La séduction idéologique de l'œuvre (Gouverneurs de la Rosée)', in Frantz-Antoine Leconte (ed.), *Jacques Roumain et Haïti: la mission du poète dans la Cité*, pp. 91–101. Paris: L'Harmattan,

______. (2012). 'Décentralisation: opportunités, limites et contraintes', in *Haïti, Réinventer l'avenir*, sous la direction de J.-D. Rainhorn, pp. 161–170. Paris: Editions MSH/Université d'Etat d'Haïti.

Figueroa, V. (2015). *Prophetic Visions of the Past. Pan-Caribbean Representations of the Haitian Revolution*. Columbus: Ohio State University Press.

Fingers, S. (2010). *Black Adorning Sounding the Call*. Leeds: Betterment.

Fleurimond, W.-K. (2003). *La communauté haïtienne en France: dix ans d'histoire (1991–2001)*. Paris: L'Harmattan.

______. (2009). *Haïti de la crise à l'occupation. Histoire d'un chaos (2000–2004)*, Tome I. Paris: L'Harmattan.

Fombrun, G. (2003). *Pour que Renaissent le Prestige & l'Espoir*. Port-au-Prince: presses de l'Imprimeur II.

Foner, N. (1978). *Jamaica Farewell: Jamaican Migrants in London*. Berkeley and Los Angeles: University of California Press.

______. (1983). *Jamaican Migrants: A Comparative Analysis of the New York and London Experience*. New York: New York University, Faculty of Arts and Science, Center for Latin American and Caribbean Studies.

______. (1986). 'Race and Color: Jamaican Migrants in New York and London', *International Migration Review*, 19, 708–727.

______. (1997). 'What's new about transnationalism? New York immigrants today and at the turn of the century', *Diaspora*, 6 (3), 355–375.

______. (1998). 'Towards a Comparative perspective on Caribbean migration' in Chamberlain (ed.), *Caribbean Migration: Globalized identities*, Routledge.

Foner, N., and Bertossi, C. (2011). 'Immigration, Incorporation, and Diversity in Western Europe and the United States: Comparative Perspectives', SAGE publications, *American Behavioral Scientist*, 55 (12), 1535–1540.

Fontaine, P-M. (1976). 'Haitian Immigrants in Boston: A Commentary', in R. Bryce-Laporte, Mortimer (eds), *Caribbean Immigration to the United States*, 1, Washington.

Fontus, F. (2001). *Les églises protestantes en Haïti. Communication et inculturation.* Chapter 3: 'L'œuvre protestante en Haïti', pp. 87–122. Paris: L'Harmattan.

Forsdick, C., and Munro, M. (2015). 'Haiti in a globalized frame', *francosphères*, vol.4, n°1, pp. 1–5.

Fouchard, J. (1972). *Langue et littérature des Aborigènes d'Ayiti*, Paris: Editions de l'Ecole.

Fraiser, N. (2016). 'Une femme à la Maison Blanche: un symbole qui ne suffit pas', *Le Monde*, 28 April, 26.

Frank, H. (2005). 'Emile Saint-Lot, Docteur Honoris Causa. Son rôle monumental dans l'indépendance de la Lybie', *L'Haïti-Observateur*, (1), 8 juin.

Francklin, Fr., and de Pazzis, E. (1997). *Paysan de Dieu. La longue route du peuple haïtien*. Paris, Bayard éditions.

Frenette, Y. (1985). *Perception et vécu du racisme par des immigrantes et des immigrants haïtiens au Québec*, Rapport CRC, Université de Montréal.

Frith, N. (2016). *Mémoire, Esclavage et Réparation*, independent report, AHRC, UK.

Frybès, M. (1991). *L'Intégration des minorités.* Paris: Gallimard.

Gaillard, G. K. (1990). *L'expérience haïtienne de la dette extérieure (1875–1915)*, Port-au-Prince: Deschamps.

Gaillard, R. (1973). *Les cents jours de Rosalvo Bobo ou une mise à mort politique*, Port-au-Prince: Presses Nationales, 269 p.

______.(1982). *Charlemagne Péralte le Caco*. Port-au-Prince: Le Natal.

______.(1983). *La guérilla de Batraville 1919–1934*. Port-au Prince: Le Natal.

Galindo, A. (2013). *The Gulf and Latin America. An Assessment of Expectations and Challenges*. Gulf Research Center Cambridge.

Gallibour, E. (2003). *Diaspora et métissage. Quelques réflexions sur les associations haïtiennes en Guyane*, communication à la 3è journée d'études doctorales sur la Caraïbe, <http://www.gensdela caraibe.org/>.

Garrigus, J. D. (2006). *Before Haiti: Race and Citizenship in French Saint-Domingue.* New York: Palgrave Macmillan.

Gayot, F. (Mgr) (2006). *L'état de la migration en Haïti*, rapport à la Commission Internationale Catholique pour la migration.

Gazan, D. (1990). *Une approche culturelle de la migration haïtienne en région parisienne durant les quinze dernières années*, Master's Thesis Ethnology, Paris VIII.

Gazeau-Secret, A. (2010). 'Francophonie et diplomatie d'influence', *Géoéconomie*, 4 (55), 39–56.

Geggus, D., and Gaspar, D. (eds) (1997). *A Turbulent Time: The French Revolution and The Greater Caribbean*. Bloomington: Indiana University Press.

Geneus, V. (2008). *Autant en emporte la Révolution. Notes sur la contribution d'Haïti à la lutte des peuples pour leur libération (Etats-Unis, Venezuela, Mexique, Cuba, République Dominicaine)*. Port-au-Prince: Bibliothèque Nationale d'Haïti – L'imprimeur ii.

______. (2014). *La Route Jamaïcaine. Une chronique des relations historiques entre Haïti et la Jamaïque*. Port-au-Prince: Bibliothèque Nationale d'Haïti – L'imprimeur ii.

George, A., and Bennett, A. (2005). *Case Studies and Theory Development in the Social Sciences*. Cambridge, MA: MIT Press.

Georges, E. (1990). *The Making of a Transnational Community: Migration, Development, and Cultural Change in the Dominican Republic*, New York: Columbia University Press.

Gilbert-Roberts, T.-A. (2013). *The Politics of Integration Caribbean Sovereignty Revisited*. Kingston: Ian Randle Publishers.

Gilroy, P. (1993). *The Black Atlantic: Modernity and Double Consciousness*. Cambridge, MA: Harvard University Press.

Girard, P. (2009). 'Rebelles with a Cause: Women in the Haitian War of Independence, 1802–1804,' *Gender & History* , 21: 60–85.

Girard, R. (2017). *Quelle Diplomatie Pour La France? Prendre les réalités telles qu'elles sont*. Paris: Les Editions Cerf.

Girault, C. (1985). 'Haïti: un pays si peu francophone et pourtant', *Herodote*, 37, 249–260.

Girod, M. (2004). *Penser le racisme. De la responsabilité des scientifiques*. Calmann-Lévy.

Girvan-Campbell, C. (2010). 'Jamaica and the Canadian Diaspora', in Delano Franklyn (ed.), *The Jamaican Diaspora: Building an Operational Framework*. Kingston: Wilson Franklyn Barnes, pp. 184–202.

Glaude, S. (2001). *Le dynamisme associatif des migrants haïtiens en France métropolitaine*, Master's Thesis, Université Paris 8, Saint-Denis.

Glick Schiller, N. (2011). 'Locality, Globality and the Popularization of a Diasporic Consciousness: Learning from the Haitian Case', in R. O. Jackson, *Geographies of the Haitian Diaspora*. New York: Routledge, Foreword.

Glick Schiller, N., Basch, L., and Szanton Blanc, C. (1992). *Towards a transnational perspective on migration: race, class, ethnicity and nationalism reconsidered*. New York: New York Academy of Sciences.

Glick Schiller, N., and Caglar, A. (2006). 'Towards a theory of locality, in migration studies: migrant incorporation and city scale', Poitiers, *Colloque* 1985–2005.

Glick Schiller, N., and Fouroun, G.-E. (2001). *Long distance nationalism and the search for home*. Durham, NC/London: Duke University Press.

Glissant, E. (1993). *Tout-Monde*. Paris: Gallimard.

Gonin, P. (1997). *D'entre deux territoires: circulations migratoires et développement entre le bassin du fleuve Sénégal et la France*, Lille, Université des sciences et technologies, HDR de géographie, 384 p.

______. (2008). 'Migrations internationales: questions de démographie et de Développement', *Migrations société*, 20 (117–118), 207–215.

______. (2010). 'International migration and territorial (re)construction: the place and role of migrants as 'frontiers runners' in development ', in C. Audebert and M. K. Doraï (eds), *Migration in a Globalised World*, IMISCOE, pp. 161–180. Amsterdam University Press.

Gonin, P., and Kotlock, N. (2009). 'Les associations de développement des migrants dans leur pays d'origine', *Accueillir*, *252*,12–16.

Gonin, P., and Lassailly-Jacob, V. (2002). 'Les réfugiés de l'environnement. Nouvelle catégorie de migrants forcés', *REMI*, 18 (2).

Gontcharoff, G. (2002). 'Petite Généalogie historique et conceptuelle du développement local', *Territoires*, 431, 5–9.

Gorgeon, C. (1985). *Gestion urbaine d'une immigration: le cas de l'Ile de Cayenne*, PhD, IAR, Université d'Aix-Marseille III.

______. (1985). 'Immigration clandestine et bidonvilles en Guyane, les Haïtiens à Cayenne', *REMI*, Poitiers, 1 (1), 143–158.

Gorgeon, C., and Calmont, R. (1987). 'L'immigration haïtienne en Guyane', *Equinoxe*, Cayenne, 23, 1–16.

Goulbourne, H. (1985). *The State, politics and violence in the Anglophone Caribbean*. Coventry: Centre for Caribbean Studies, University of Warwick, the 1st Walter Rodney memorial lecture.

______. (1999). 'Exodus? Some social and policy implications of return migration from the UK to the commonwealth Caribbean in the 1990s', *Policy Studies* 20, n° 3, pp. 157–172. *Knowledge*, VII, 4, Fall, pp. 51–72.

______. (2002). *Caribbean Transnational Experience*. London: Pluto Press, 246 p.

______. (2002). 'Familles caribéennes en Grande-Bretagne', *Diasporas caribéennes, Hommes et migration*, n° 1237, pp. 70–81.

______.(2009). 'Forced and Free Caribbean Migration: An Understanding of Modern Diasporas', in Elizabeth Thomas-Hope (ed.), *Freedom and Constraint in Caribbean Migration and Diaspora*. Ian Randle Publ., 392 p.

Goulbourne, H., and Solomos, J. (2003). 'Families, Ethnicity, and Social Capital', *Social Policy and Society*, 2 (4), 329–338.

Gouraige, G. (1974). *La Diaspora des Haïtiens en Afrique*. Sherbrooke: Editions Naaman.

Granovetter, M. (1973). 'The Strength of the Weak Ties', *American Journal of Sociology*, 78 (6), 1360–1380.

Grasmuck, S. (1982). 'Migration within the Periphery: Haitian Labor in the Dominican Sugar and Coffee Industries', *IMR*, 16 (2), 365–377.

Grawitz, M. (1993). *Méthodes des sciences socials*. Paris: Dalloz.

Green, A. (2007). 'Birmingham, Equiano and the Transatlantic Slave Trade', *Equiano Enslavement, Resistance & Abolition*, pp. 42–49, Birmingham Museum.

Green, Cecilia A. (2015). 'The New Chinese Presence in the Caribbean: Towards a Global Understanding', in *Beyond Free Trade, Alternative Approaches to Trade, Politics and Power*, pp. 97–115, London: Palgrave MacMillan.

Green, N. (2002). *Repenser les migrations*, Paris: PUF, 144 p.Grégoire, N. (2011). 'Émigration des ressources humaines qualifiées en provenance des pays de la Caraïbe et de l'Amérique latine: le cas d'Haïti', in Joint, L-A. and Mérion, J. (eds), *L'immigration haïtienne dans la Caraïbe*,pp. 63–68. Gourbeyre: Éditions Nestor,

Griggs, E. L., and C. H. Prator (eds) (1968). *Henry Christophe and Thomas Clarkson: A Correspondence*. New York: Greenwood Press.

Grosfoguel, R. (1998). 'Modes of incorporation: colonial Caribbean migrants in Western Europe and the United States', in Mary Chamberlain, *Caribbean Migration: Globalized Identities.* USA and Canada: Routledge.

______. (2007). 'Racisme culturel et immigrés caraïbéens dans les métropoles de l'économie monde capitaliste', in *Dynamiques migratoires de la Caraïbe* (coll. Terres d'Amérique), pp. 105–127. Karthala/Géode Caraïbe.

Guarnizo, L. E. (2003). 'The Economics of Transnational Living', *International Migration Review*, 37 (3), 666–699.

Guarnizo, L. E., and Smith, M. P. (1998). *Transnationalism from Below*. New Brunswick, NJ/London: Transaction Publishers.

Gubert, F. (2005). 'L'impact des transferts de fonds sur le développement des pays d'origine: le cas de l'Afrique', *Migrations, transferts de fonds et développement*, Editions OCDE, pp. 43–72.

______. (2007). 'Migrations et transferts de fonds. Impact sur les pays d'origine', *Revue d'économie du développement*, 21 (3), 183–188.

Guengant, J.-P. (1988). 'L'exil haïtien', *Antiane*, Pointe-à-Pitre (6), 16–19.

______. (1996). 'Migrations internationales et développement: les nouveaux paradigmes', *REMI,* 12 (2).

Guérin, D. (1956). *Les Antilles Décolonisées*. Paris: Présence africaine.

Guilmoto, C. Z., and Sandron, F. (2003). *Migration et développement*. Paris: La Documentation Française.

Haas, H. de (2005). 'International Migration, Remittances and Development: Myths and facts', *Third World Quarterly* 26 (28): 1269–1284.

______. (2010). 'Migration and Development: A Theoretical Perspective', *International Migration Review* 44 (1): 227–264.

Hadjadj, B. (2007). *L'an prochain à Port-au-Prince. Sortir de l'esclavage.*Paris: Maisonneuve & Larose.

Halbwachs, M. (1992). *On Collective Memory.* Chicago: University of Chicago Press.

Hall, C. (2002). *Civilizing Subjects. Metropole and Colony in the English Imagination, 1830–1867.* Cambridge: Polity Press.

Hall, E. T. (1969). *The Hidden Dimension.* New York: Anchor Books.

Hall, S. (1991). 'Old and New Identities, Old and New Ethnicities', in Anthony King (ed.), *Culture, Globalization, and the World System*, 41–68. Basingstoke: MacMillan.

Hamidi, C. (2002). *Les effets politiques de l'engagement associatif. Le cas des Associations issues de l'immigration*, Thèse doctorale, IEP de Paris, 747 p.

Hamilton, J. M. (1983). *Elliptical Sculptured Grindings Stones of Hispaniola.* St Augustine: University of Florida, 23 p.

Handerson, J. (2012). *Inauguration of a New Era in the XXI Century Migration: Fomasyon diaspora ayisyen lot bodlo-Brezil*, Paper (November) HSA Annual Conference, CUNY, New York.

Harrington, A., Prest, S., and Unheim, P. (2009). 'Jamaica, Haiti, and the Role of Diasporas', *Human Architecture: Journal of the Sociology of Self-Knowledge*, 7 (4), 15–36.

Harris, C. (1993). 'Post-war Migration and the Industrial Reserve Army', in J. Winston and C. Harris (eds). *Inside Babylon: The Caribbean Diaspora in Britain*, pp. 9–51. London: Verso.

Hart, D. W. (2017). 'The Globalectics of Nationalism in C. L. R. James's and Michelle Cliff's Writing', *Postcolonial Text*, Vol 12, n°2.

Hector, C. (2012). 'Quelques perspectives des rapports "paradiplomatiques" entre la Prusse de Friedrich Wilhem III et le Royaume d'Henry Christophe (1811–1820)', in *L'historiographie allemande et la Révolution haïtienne, Approches récentes*, n°245/8.

______. (2016). *Daniel Fignolé et le rassemblement fignoliste du nord (RFN): un apport méconnu dans la crise politique de 1956–1957.* Montreal: Les éditions du CIDIHCA.

Hector, M., and Hurbon, L. (eds) (2009). *Genèse de l'Etat haïtien (1804–1859)*, Fondation MSH – Presses Nationales d'Haïti, Paris, Port-au-Prince.

Heine, J., and Manigat, L. F. (eds) (1988). *The Caribbean and World Politics: Cross Currents and Cleavages.* New York: Holmes & Meier, 404 p.

Helly, D., Labelle, M., and Larose, S. (1980). 'L'émigration haïtienne, un problème national', *Collectif Paroles*, Montréal, 2, 18–26.

Henke, H. (1996a). *Dependency and Foreign Relations: A Comparative Study of the Manley and Seaga Governments in Jamaica, 1972–1989*, PhD, UWI, Mona.

______. (1996b). 'Mapping the "Inner Plantation": A Cultural Exploration of the Origins of Caribbean Local Discourse', *Social and Economic Studies*, 4 (45), 51–75.

______. (2001). *The West Indian Americans*. Wesport, CT: Greenwood Press.

______. (2003). 'Freedom Ossified. Political Culture and the Public Use of History in Jamaica', in Henke and Reno (eds), *Modern Political Culture in the Caribbean*, pp. 111–140. Mona: University of the West Indies Press.

Henke, H., and Reno, F. (eds) (2003). *Modern Political Culture in the Caribbean*. Kingston: University of the West Indies Press.

Henri, R. (1960). 'Les traffics coloniaux du port de la Rochelle au XVIIIème siècle', in *Mémoire de la Société des Antiquaires de l'Ouest*, 4e. série, IV, Ix-209.

Heran, F. (2017). *Avec L'immigration. Mesurer, débattre, agir*. Paris: éditions La Découverte.

Hermantin, F. (2011). 'Droit de l'immigration, pratiques administratives et Judiciaires', L.-A. Joint et J. Mérion (eds), *L'immigration haïtienne dans la Caraïbe*, Gourbeyre (Guadeloupe): Éditions Nestor, pp. 97–110.

Hidair, I. (2005). 'Impact de la croissance démographique sur les actions du Conseil Général de la Guyane. Le poids de l'immigration', Cayenne.

Hobsbawn, E. J. (1992). *Nations and Nationalism Since 1780: Programme, Myth, Reality*, Cambridge: Cambridge University Press.

Hoffman, J-F L. (1993). 'Un mythe national: la cérémonie du Bois-Caïman', in *La République Haïtienne: Etats des lieux et perspectives*, Paris, ADEC, Karthala, 434–448.

Hollifield, J. (1990). 'Migrants ou citoyens: la politique de l'immigration en France et aux Etats-Unis', *REMI*, 6 (1), 159–177.

Houée, P. (1990). 'Le développement local: imbrication d'acteurs et de dynamiques sociales', *Cahiers du LERASS*, 20, 15–19.

______. (2001). *Le développement local au défi de la mondialisation*. Paris: L'Harmattan.

Hume, Y. (2011). 'On the Margins: The Emergence of a Haitian Diasporic Enclave in Eastern Cuba', in R. Jackson (ed.), *Geographies of the Haitian Diaspora*. pp. 71–90. New York: Routledge,

Hurbon, L. (1982a). 'La fuite du peuple haïtien ou les causes de la migration actuelle', *Collectif Paroles*, Montréal, 16, 34–37.

______. (1982b). 'La migration haïtienne en Guadeloupe', *Collectif Paroles*. Montréal, 16, 26–27.

______. (1983). 'Racisme et sous-produit du racisme: immigrés haïtiens et dominicains en Guadeloupe', *Les Temps Modernes*, 1983, n° 441–442.

______. (2001). *Pour une sociologie d'Haïti au XXIe*, Paris: Editions Karthala, 301 p.

______. (2004). *Religions et lien social. L'Eglise et l'Etat moderne en Haïti*, Paris: Editions du Cerf, 317 p.

Icart, Jean Claude (1984). *Essai sur les boat-people haïtiens en Floride*, Montréal, CIDIHCA, 188 p.

______. (1987). *Négriers d'eux-mêmes. Essai sur les boat people haïtiens en Floride*, Montréal, CIDIHCA, 188 p.

Institution Sainte Rose de Lima, 'Présence des Sœurs de Saint Joseph de Cluny en Haïti et dans le monde', Port-au-Prince, 36 p.

Itzigsohn, J. (2000). 'Immigration and the Boundaries of Citizenship: The Institutions of Immigrants' Political Transnationalism', *IMR*, 34, 1126–1154.

Jackson, R. O. (ed.) (2011). *Geographies of the Haitian Diaspora*. New York: Routledge.

Jacquot, E. (2010). *Les Spiritains en Haïti (1843–2003) d'Eugène Tisserant (1814–1845) à Antoine Adrien (1922–2003)*. Paris: Karthala.

James, C. L. R. (2008). *Les Jacobins Noirs Toussaint Louverture et la Révolution de Saint-Domingue* (réédition 1938, 1963, 1980, 2001). Paris: éditions Amsterdam.

James, C., and Perivolaris, J. (2000). *The Cultures of the Hispanic Caribbean*. Gainesville: University of Florida Press.

Janis, W. (2004). *Haitian Odyssey & the Jamaican Connection*, Kingston: Pelican Publishers.

Javouhey, Anne-Marie, fondatrice de la congrégation des Sœurs de St-Joseph de Cluny, *Correspondance*, t.1 (1798–1833), Texte présenté par les Sœurs Jean Hebert et Marie-Cécile de Segonzac (1994), Paris: Cerf, 559 p.

Jean, G. (2002). *Sociétés civiles en mutation – Haïti*, CIPA/Centre International de Politologie Appliquée, Port-au-Prince.

Jeffrey, C. (2013). Geographies of children and youth III: Alchemists of the revolution? *Progress in Human Geography*, 37(1), 145–152.

Jeffry, D. (2011). 'L'immigration haïtienne vue de Saint-Martin', L.-A. Joint and J. Mérion (eds), *L'immigration haïtienne dans la Caraïbe*, pp. 163–186. Gourbeyre: Éditions Nestor.

Joachim, B. (1971). 'Le néocolonialisme à l'essai. La France et l'indépendance d'Haïti', *La Pensée*, 15: 48 .

______. (1979). *Les racines du sous-développement en Haïti*. Port-au-Prince: Deschamps.

Joint, L.-A. (2005). *Système éducatif et inégalités sociales en Haïti. Le cas des écoles catholiques congréganistes Saint Martial, Saint-Louis de Bourdon et Juvénat du Sacré Cœur*, PhD Thesis. Paris: EHESS.

______. (2006). *Système éducatif et inégalités sociales en Haïti*, L'Harmattan.

______. (2012). *L'insertion des immigrés haïtiens dans la société guadeloupéenne: quelles initiatives et quels obstacles ?*, CSA Conference.

Joint, L.-A., and Mérion, J. (eds) (2011). *L'immigration haïtienne dans la Caraïbe*. Gourbeyre: Éditions Nestor.

Joly, D. (1987). 'Associations amongst the Pakistani population in Britain', in J. Rex, D. Joly and C. Wilpert (eds), *Immigrant Associations in Europe*, pp. 62–85. Aldershot: Algate Publishing Ltd.

______. (1990). 'Etude comparative France-Grande Bretagne', coll. D. Lapeyronnie et al., *L'intégration des Minorités Immigrées*, pp. 345–387. Paris.

______. (1992). 'Minorités ethniques et risques de ségrégation', in D. Lapeyronnie (ed.), *Immigrés en Europe Politiques locales d'intégration*, pp. 111–136. Paris: La documentation Française,

______. (1996). *Heaven or Hell ? Asylum Policies and Refugees in Europe*. London: MacMillan Press Ltd.

______. (2001). *Blacks and Brittany*. Aldershot: Ashgate Publishing Limited.

Joly, D., and W. Khursheed (2012). 'Migration Policy and Irregular Workers in the United Kingdom', *Note de l'Ifri Center for Migrations & Citizenship.*

Jones, T.-A. (2008). 'L'intégration des immigrants jamaïcains à Miami et Toronto', *REMI*, 24 (1), 139–158.

Joppke, C. (2001). 'The Evolution of Alien Rights in the United States, Germany and the EU', in A. Aleinikoff, D. Klusmeyer (eds), *Citizenship Today: Global Perspectives and Practices*. pp. 36–62. Washington: Carnegie, Endowment for International Peace.

______. (2007). 'Transformation of Citizenship: Status, Rights, Identity', *Citizenship Studies*, Routledge, 11 (1), 37–48.

Joseph, A. Lafontant (2000). *Décentralisation en Haïti: réforme de la législation et de la politique de décentralisation*, Port-au-Prince: édition Grafopub.

Joseph, P. (2004). *Dictionnaire historique et géographique des communes d'Haïti*. Québec: Editions Combit.

Josselin, J. (2004). *L'insertion des migrants haïtiens en Guadeloupe*, Master's (Geography), Université des Antilles et de la Guyane.

Kastoryano, R. (1997). 'Participation transnationale et citoyenneté: les immigrés dans l'Union européenne', *Cultures & Conflits*, 28 (Winter).

______. (1998). 'Transnational participation and citizenship: Immigrants in the EU', Transnational Communities Paper series, Oxford, Oxford University Press.

______.(2006). 'Vers un nationalisme transnational. Redéfinir la nation, le nationalisme et le territoire', *Revue française de science politique*, Presses de Sciences Po, 56 (4), 533–553.

Katz, J. M. (2013). *The Big Truck That Went By: How the World Came to Save Haiti and Left Behind a Disaster*. Palgrave Macmillan.

______. (2015). 'The King and Queen of Haiti', *Politico Magazine*, 4 May.

Kempeners, M., and Neill, G. (1985). *Trajectoires socio-professionnelles des immigrantes et immigrants haïtiens au Québec*. Université de Montréal.

Kivisto, P. (2001). 'Theorizing transnational immigration: a critical review of current efforts', *Ethnic and Racial Studies*, 24, 549–577.

Klein, J.-F., Singaravélou, P., and de Suremain, M.-A. (2012). *Atlas desempires coloniaux XIXe-XXe siècles*. Paris: Éditions Autrement.

Knight, F., and Palmer, C. A. (1989). *The Modern Caribbean*. Durham, NC: UNC Press Books.

Kotlok, N. (2005). 'Emigration et Insularisme au Cap Vert', in C. Charef and P. Gonin (eds), *Emigrés-Immigrés dans le développement local*, pp. 59–72. Agadir: Ormes/ Sud Contact.

Kotlok, N., and Gonin, P. (2007). 'Les territorialités migrantes à l'épreuve des projets de développement: le Guidimakha mauritanien', XLIIIème colloque de l'ASRDLF Grenoble-Chambéry.

Kotlok, N., and Semedo, C. (2009). 'Les territorialités migrantes à l'épreuve des projets de développement: le partenariat Aubervilliers-Boully', in *Migrants d'Ici et d'Ailleurs Du transnational au Local*, K. Akoka, P. Gonin et D. Hamelin (eds), Atlantique.

Kritzler, E. (2008). *Les pirates juifs des Caraïbes*. André Versailles éditeur.

Kuyu, C. (2006). *Les Haïtiens au Congo*. Paris: L'Harmattan.

Kymlicka, W. (2002). *Contemporary Political Philosophy An Introduction*, Oxford: Oxford University Press, 497 p.

______. (2004). 'Le mythe de la citoyenneté transnationale', *Critique internationale* 2 (23), 97–111.

La Grenade-Lashley, M. (2016). *Mwen Ka Alé. The French-Lexicon Creole of Grenada History, Language and Culture*. Chula Vista, CA: Aventine Press.

Laaroussi, M. V. (2008). 'Immigration en région: le territoire local à l'épreuve de la mobilité et des réseaux transnationaux', in X. Leloup (ed.), *Les nouveaux territoires de l'ethnicité*, pp. 79–105.Québec: les presses de l'université Laval.

Labelle, M., et al. (2001). *Incorporation citoyenne et/ou exclusion. 2ème génération issue de l'immigration haïtienne et jamaïcaine*, Montréal, CRIEC.

Labelle, M., Larose, S., and Piché, V. (1983). 'Émigration et immigration: les Haïtiens au Québec', *Sociologie et sociétés*, Montreal, 15 (2).

Labelle, M., and Midy, F. (1999). 'Re-reading citizenship and the Transnational practices of immigrants', *JEMS*, 25 (2), 213–232.

Labelle, M., and Rocher, F. (2004). *Contestation transnationale, diversité et citoyenneté dans l'espace québécois*. Presses Université Québec.

Lacoste, Y. (1985). *La géographie ça sert d'abord à faire la guerre*. Paris: La Découverte.

______. (1989). *Géographie du sous-développement*, Paris: PUF.

______.(1991). 'Les territoires de la nation', *Hérodote*, 62–63, 3ème T, 1–21.

______.(2006). *Géopolitique. La longue histoire d'aujourd'hui*. Paris: Larousse.

Lacroix, T. (2003). *Espace transnational et territoires. Les réseaux marocains du développement*, PhD Thesis, Université de Poitiers.

______. (2005). *Les réseaux marocains du développement. Géographie du transnational et politiques du territorial.* Paris: Sciences Po.

______. (2008). 'Les Marocains et Sénégalais de France, permanences et évolutions des relations transnationales', *REMI*, 24 (2), 23–43.

______. (2009). Migration, développement, Codéveloppement: quels acteurs pour quels discours ? *Rapport de synthèse (IDEM)*, Institut Panos Paris.

______. (2010). Hometown Organizations and Development Practices, *IMI* Oxford, Working Papers, Paper 28 (Dec.).

Laëthier, M. (2007). *Être haïtien et migrant en Guyane française*, PhD Thesis (Anthropology), École des hautes études en sciences sociales (EHESS).

______. (2009). 'Les associations haïtiennes sur la scène multiculturelle Guyanaise', *Congrès AFSP*.

Lafleur, J.-M. (2005). *Le transnationalisme politique: pouvoir des communautés immigrées dans leurs pays d'accueil et d'origine*, Louvain.

______. (2008). *Political Transnationalism and the State*, PhD Thesis IEP/ Liège.

Laguerre, M. (1984). *American Odissey: Haitians in New York City*. Ithaca, NY: Cornell University Press.

______. (1995). 'Le circuit transnational: le champ multinational des pratiques politiques de la diaspora haïtienne aux Etats-Unis', in G. Brochu, M. Labelle, J. Lévy (eds), *Etudes et recherches sur la pluriethnicité*. Université du Québec, pp. 1–16.

______. (1998). *Diasporic Citizenship: Haitian Americans in Transnational America*, New York: Macmillan Press Ltd, 222 p.

______. (2009). 'The Transglobal Network Nation: Diaspora, Homeland, and Hostland', in E. Ben-Rafael et al. (eds), *Transnationalism: Diasporas and the Advent of a New (dis)Order*, Leiden: Brill Academic Publishing.

______. (2013). *Parliament and Diaspora in Europe*. New York: Palgrave Macmillan.

Lahusen, S. (1986). 'Retours à Haïti', *Réfugiés* (29), 21–22.

Lajarge, R. (2008). 'Pas de territorialisation sans action (et vice versa ?)', in M. Vanier (ed.), *Le Pouvoir des territoires Essai sur l'interterritorialité*, pp. 193–209. Paris: Economica,.

Lalla, B., et al. (2013). *Methods in Caribbean Research Literature, Discourse, Culture*. University of the West Indies Press.

Lamaute, A. (1999). *La bourgeoisie nationale: une entité controversée*. Montréal: Les Editions du CIDIHCA.

Lambourdière, Éric (ed.) (2007). *Les Caraïbes dans la géopolitique mondiale*. Paris: Ellipses Editions (coll. Carrefours).

Landes, D. S. (1999). *Wealth and Poverty, Why Some are So Rich and Some are So Poor*. New York: W. W. Norton & Company.

Laperriere, A. (1998). ' Dépasser le racisme ? L'expérience contrastée de jeunes Montréalais d'origine haïtienne'. *REMI*, 14 (1), 121–139.

Lapeyronnie, D. (1993). *L'individu et les minorités. La France et la Grande-Bretagne face à leurs immigrés*. Paris: PUF.

Lara, O. D. (1992). *Caraïbes en construction: espace, colonisation, résistance*. Paris: Éditons du CERCAM.

______. (2004). 'Le rayonnement de la Révolution haïtienne aux Caraïbes-Amériques', in *Luttes contre l'esclavage*, édition spéciale *UNESCO*, 12–15.

Larose, S. (1984). 'Transnationalité et réseaux migratoires: entre le Québec, les Etats-Unis et Haïti', *Les cahiers de recherche sociologique* (2), 115.

Laurent, E., et al. (2017). 'After Brexit ... Securing Economic Interests', *The Ramphal Institute*. London: Rila Publications Ltd

Layton-Henry, Z. (2007). *Transnational Communities, Citizenship and African-Caribbeans in Birmingham*, <http://www.transcomm.ox.ac.uk/working%20 papers/WPTC-02-07%20LaytonHenry.pdf>.

Le Glaunec, J.-P. (2014). 'L'armée indigène. La défaite de Napoléon en Haïti'. Préface de Lyonel Trouillot. Québec: Lux Editeur.

Lebon, A. (2001). *Immigration et présence étrangère en France en 1999*. Paris: La Documentation française.

Leca, J. (1986). Individualisme et citoyenneté, in J. Leca et P. Birnbaum, *Sur l'individualisme*, pp. 159–209. Paris: Presses de la FNSP.

______. (1992). 'Nationalité et citoyenneté dans l'Europe des immigrations', in J. Costa-Lascoux and W. Patrick (eds), *Logiques d'Etat et immigrations*, pp. 13–57. Paris: Kimé.

Lechaume, A. (2001). 'Repenser la Caraïbe: constructions culturelles et identité territoriale', in P. Guillaume (ed.), *Identités caraïbes*, pp. 269–273. Paris: Éditions du Comité des travaux historiques et scientifiques

Leconte, V. (2015). *Henri Christophe dans l'Histoire d'Haïti*, Port-au-Prince: éditions Fardin.

Lemay-Hebert, N. (2010). 'L'engagement international en Haïti à la lumière des leçons des missions onusiennes à Kosovo et au Timor-Leste', in *Entre Refondation et reconstruction: les problématiques de l'avenir post-sismique d'Haïti*, sous la direction de W. Dorlus, Editions de l'université d'Etat d'Haïti, pp. 103–127.

Lemoine, M. (1981). *Sucre amer. Esclaves aujourd'hui dans les Caraïbes*. Paris: Encre.

Leveau, R., and Wihtol de Wenden, C. (2001). *La 'Beurgeoisie'. Les trois âges de la vie associative issue de l'immigration*, Paris: CNRS Editions.

Lévêque, K. (1970). *De la théologie politique à la théologie de la révolution*, Frères du Monde, No. 64, 2, 1970, p. 26 à 40.

______. (1971). 'L'interpellation mystique dans le discours duvalérien', in *Nouvelle Optique*, 1, n° 4, pp. 5–32.

______. (1986). 'Impressions d'un retour d'exil', *Relations*, texte écrit le 14 mars, p. 77–78.

Levison, A. C. (1987). *La migration haïtienne au Vénézuéla, un cas d'étude de l'insertion à l'intégration*, PhD. Thesis Paris III Sorbonne nouvelle, IHEAL.

Levitt, P. (2001). *The Transnational Villagers*, Los Angeles: University of California Press.

______. (2009). *God needs no passport: Immigrants and the Changing American Religious Landscape*, New York: The New Press.

______. (2010). 'Taking Culture Seriously: the Unexplored Nexus between Migration, Incorporation and Development', *REMI*, 26 (2), 139–153.

Levitt, P., and Glick-Schiller, N. (2004). 'Conceptualizing Simultaneity: A Transnational Social Field Perspective on Society', *IMR*, 38 (3)1002/39.

Lévy, J., and Lussault, M. (2003). *Dictionnaire de la géographie et de l'espace des sociétés*, Paris: Belin.

Lewando-Hundt, G. (2010). *Applied Social Research in Social Work and Social Policy*, Module at the University of Warwick, Fall.

Lewis, R. (2011). '"Mirror Miror on the Wall/ Who is the Fairest of Them All?" Rex Nettelford and the Knotty Issue of Identity', *Caribbean Quarterly*, 57 (3/4), 33–48.

Lewis, V. A. (1984). 'Geopolitical Realities in the Caribbean', in Elizabeth Thomas-Hope (ed.), *Perspectives on Caribbean Regional Identity*, pp. 5–15, Liverpool: CLAS.

Lindskog, D. P. (1998). 'From Saint-Domingue to Haiti: some consequences of European colonisation on the physical environment of Hispaniola', *Caribbean Geography*, 9 (2), 71–86.

Lilite, O. (2004). Conception cartographique de cartes de pauvreté d'Haïti, Unité de Télédétection de Systèmes d'Information Géographique (UTSIG), Ministère de la Planification (MPCE) ; CATESIG (Université Quisqueya).

Lima, S. (1998). *Mobilité Recomposition Territoriale et Décentralisation au Mali. Le cas de la région de Kayes*, DEA Geography, Montpellier.

______. (2003a). *Découpage entre espace et territoire: la fin des limites ? La fabrique desterritoires communaux dans la Région de Kayes, Mali*, Thèse, Université de Poitiers.

______. (2003b). 'Les espaces associatifs face aux communes rurales Recompositions Spatiales dans la Région de Kayes (Mali)', in M. Charef and Gonin (eds), *Emigrés/immigrés*, pp. 285–309. Agadir: Ed Sud.

______. (2007). 'Le dépassement des territoires Bâtisseurs et passeurs et d'espaces', in *Territoires, territorialité, territorialisation: et après ?*, Colloque interdisciplinaire et international, Grenoble (juin), 7 p.

______. (2008). 'Penser l'espace autrement entre proximité sociale et distance spatiale', in *Territoires en action et dans l'action*. Rennes: PUR.

______. (2009). Repenser les territoires du politique. Eléments de réflexion à partir de la décentralisation au Mali, *CUFR*, 15 p.

Livesay, D. (2018). *Children of Uncertain Fortune: Mxed-Race Jamaicans in Britain and the Atlantic Family, 1733–1833*, Omohundro Institute and University of North Carolina Press.

Lloyd, C. (1993). 'L'expérience de la Grande Bretagne', in C. Wihtol de Wenden and O. Grandmaison Le Cour (eds), *Les étrangers dans la cité expériences européennes*, pp. 169–175. Paris: La Découverte.

Lorcerie, F. (2007). 'Le primordialisme français, ses voies, ses fièvres', in M.-C. Smouts (ed.), *La situation post-coloniale. Les post colonial studies dans le débat français*. Paris: Presses de Sciences Po.

Louidor, W. E. (2011). 'L'Amérique du Sud: pôle d'attraction de la migration haïtienne', *Vivre Ensemble*, Centre Justice et Foi, 19 (63), 1–5.

Louis Martin, M. (1994). 'La France dans la Caraïbe: la fonction géopolitique des départements français d'Amérique', in M. Burac (ed.), *Guadeloupe, Martinique et Guyane dans le monde américain*, pp. 247–273. Karthala.

Lowenthal, D. (1994). 'Identity, heritage, and history', in J. R. Gillis (ed.), *Commemorations the Politics of National Identity*. Princeton, NJ: Princeton University Press, 41–57.

Loza, L. (2007). 'Les Jamaïcaines dans le sud de la Floride: exemple d'une intégration socioculturelle réussie', in *Dynamiques migratoires de la Caraïbe* (coll. Terres d'Amérique), pp. 129–139. Karthala/Géode Caraïbe.

Lucas, R., and Chappell, L. (2009). 'Measuring Migration's Development Impacts: Preliminary evidence from Jamaica.' *Development on the Move: Working Paper 2*, London: Institute for Public Policy Research.

Luzincourt, A. F. (2005). *Les enfants des migrants haïtiens à l'école*. Paris: L'Harmattan.

Ma Mung, E. (1992). 'Dispositif économique et ressources spatiales: éléments d'une économie de diaspora', *REMI*, 8 (3), 175–193.

______. (1994). 'L'entreprenariat ethnique', *Sociologie du travail* (2), 185–210.

______. (1999). 'La Dispersion comme ressource', *Culture & Conflits* (33), 89–103.

______. (2009). 'La politique des Etats de départ vis-à-vis de leurs ressortissants à l'étranger est-elle un instrument de leur politique étrangère ?' in M. Desse and P. Gonin (eds), *Rapport de synthèse Initiatives socio-économiques des Marocains résidents à l'étranger dans les sociétés méditerranéennes*, pp. 51–56. AI n° MA/06/158.

______. (2015). 'Migrants of Chinese Origin in France: Economic and Social Integration', in L. Baldassar et al. (eds), *Chinese Migration to Europe*, MacMillan, Palgrave: pp. 49–54.

Mabanckou, A. (2005). *Ver Cassé*. Paris: Seuil.

______ (2018). 'Francophonie, langue française: lettre ouverte à Emmanuel Macron', *BIBLIOBS*, 15/01.

Mabanckou, A., and Mbembe, A. (2018). 'Le français, notre bien commun ?' *BIBLIOBS*, 12/02.

McLeod, M., Owen, D. and Khamis, C. (2001). *Black and Minority Ethnic Voluntary and Community Organisations*, London Policy Study Institute.

McLeod, M. C. (2000). *Undesirable Aliens: Haitian British West Indian Immigrant Workers in Cuba, 1898 to 1940*, PhD Dissertation, University of Texas.

Macron, E. (2016). *Révolution. C'est notre combat pour la France*. XO Editions.

Maesschalck, M. (2013). 'Subjectivation et transformation sociale. Critique du renouveau en théorie de l'action à partir de Karl Lévêque, Etienne Balibar et Louis Althusser', in *Les Carnets du Centre de Philosophie du Droit*, n° 158.

Mafundikwa, K. (2014). *The Price of Memory*, film production, Gandanga Media LLC.

Maignan, F. (2007). *Fond-des-Blancs et ses grandes familles traditionnelles*. Coconut Creek, FL: Educa Vision.

Maldonado-Torres, N. (2007). 'Thinking through the Decolonial Turn: Post-continental Interventions in Theory, Philosophy, and Critique – An Introduction', *Transmodernity: Journal of Peripheral Cultural Production of the Luso-Hispanic World* 1.1 (2011): 1–30.

Manigat, L. F. (1982). 'Existe-t-il une identité culturelle commune à la Caraïbe Insulaire', in *Problems of Caribbean Development*, Munich.

______. (2001). 'Le roi Henry Christophe et l'éducation nationale 1807–1820', in *Eventail d'histoire vivante*, tome 1, Port-au-Prince, Coll. CHUDAC: 293–309.

______. (2007). Éventail d'Histoire Vivante d'Haïti. Des préludes à la Révolution de Saint-Domingue jusqu'à nos jours (1791–2007)', t.5: *Aspects et problèmes de l'histoire de la diplomatie et des relations internationales*. pp. 459–1011. Port-au-Prince: Collection du CHUDAC.

______. (2010). 'Le rôle des intellectuels et des professionnels du savoir haïtien par rapport au pouvoir et à l'avoir', in *Entre Refondation et reconstruction: les problématiques de l'avenir post-sismique d'Haïti*, sous la direction de W. Dorlus, Editions de l'université d'Etat d'Haïti, pp. 33–90.

Manigat, M. (1988a). 'Brazil and the Caribbean', in Jorge Heine and Leslie F. Manigat (eds), *The Caribbean and World Politics: Cross Currents and Cleavages*, pp. 262–275. New York: Holmes & Meier.

______. (1988b). 'Caricom at Ten', in Jorge Heine and Leslie F. Manigat (eds), *The Caribbean and World Politics: Cross Currents and Cleavages*, pp. 94–109. New York: Holmes & Meier.

Manley, B. (2008). *The Manley Memoirs*. Kingston: Ian Randle Publishers.

Manley, Bob. (1980). 'Could you be loved', Album *Uprising*, Label Islands.
Manley, M. (1991). *The Poverty of Nations Reflections on Underdevelopment and the World Economy*, London: Pluto Press.
Marroncle, A. (2017). 'La Rochelle célèbre la Réforme sous le signe de l'œcuménisme', *La Croix*, 18 August, 11.
Marshall, D. (1982). 'The History of Caribbean Migrations: the Case of the West Indies', *Caribbean Review*, 11 (1), 6–9; 52/53.
Martin, P., Martin, S., and Weil, P. (2002). 'Best practices Options: Mali', *International Migration*, vol. 40 (3), 87–98.
Martinetti, B. (2017). *La traite négrière à La Rochelle*. La Crèche: La Geste.
Masse, R. (1983). *L'émergence de l'ethnicité haïtienne au Québec*, PhD Thesis (Sociology), Université Laval (Canada).
Massey, D., et al. (1995). *A Place in the World: Places, Cultures and Globalization*. Oxford: Open University Press.
Massey, D., et al. (1998). *World in Motion. Understanding International Migration at the End of the Millenium*. Oxford: Oxford University Press.
Mathurin, G. (2008). 'Origine de la philosophie de la révolution haïtienne', Port-au-Prince.
Maurouard, E. (2013). *Des femmes dans l'émancipation des peuples noirs de Saint-Domingue au Dahomey*. Paris: Éditions du Cygne.
Mbembe, A. (1999). 'Decolonizing Knowledge and the Question of the Archive', public lecture, Wits Institute for Social and Economic Research (WISER).______ (2017). *Critique of Black Reason*. Durham, NC: Duke University Press.
Mbiafu Mfaboum, E. (2004). 'Haïti et l'Afrique: douleurs des destinées, spécularité des douleurs', *Africultures* (janvier-mars), n° 58, p. 153.
Médam, A. (1993). 'Diaspora / Diasporas: Archétype et typologie'. *REMI*, 9 (1), 59–66.
Meddi, A. (2018). 'La France doit s'affranchir de son ethnocentrisme', *La Croix*, 20/03, 4 p.
Memmi, A. (2004). *Portrait du décolonisé*. Paris: Editions Gallimard.
Mérion, J. (2011). 'Haïti et le reste de la Caraïbe une relation paradoxale', in L. A. Joint et J. Mérion (eds), *L'immigration haïtienne dans la Caraïbe quel défi pour l'unité des peuples ?* Guadeloupe: Editions Nestor Metellus, J. (2003). *Haïti une nation pathétique*. Maisonneuve & Larose.
Meylion-Reinette, S. (2008). *De la diaspora haïtienne à la communauté haïtiano-américaine de NY: modèle d'une intégration réussie?* PhD Thesis, UAG.
Michel, G. (2003). 'Haïti: deux siècles de liberté', *Haïtiens Aujourd'hui*, revue mensuelle avril, pp. 22–23.
Midy, F. (1982). 'L'exode populaire haïtien: retraite des marrons', *Collectif Paroles*, Montréal, 16, 32–33.

Migerel, H. (1987). *La Migration des Zombis, Survivances de la magie antillaise en France*. Paris: Editions caraïbéennes.

Mignolo, W. (1993). 'Colonial and Postcolonial Discourse: Cultural Critique or Academic Colonialism?' *Latin American Research Review*, Vol. 28, No. 3, pp. 120–134.

Miles, M. A., and Huberman, M. (2003). *Analyse des données qualitatives*, 2e édition. Bruxelles: De Boec.

Millet, D., and Mauger, F. (2004). *La Jamaïque dans l'étau du FMI: La dette expliquée aux amateurs de reggae, aux fumeurs de joints et aux autres*. Paris: L'Esprit Frappeur.

Minassian, G. (2006). 'Pour une géopolitique des Diasporas', *AFRI* 7,77–89.

Mitchell, D. (2000). *Cultural Geography. A Critical Introduction*. Oxford: Blackwell Publishers.

Montas, R. (2012). 'L'économie haïtienne 1986–2011 Ouverture par la migration et stagnation', revue haïtienne de société et de culture, *Rencontre*, CRESFED, 26–27, pp. 29–35.

Montauban, G. (2004). *La vie à la Colline d'Aquin* (Haïti). Paris: Dauphin Noir.

Monza, R. (2009). 'Géopolitique de la crise guadeloupéenne: crise sociale et/ou postcoloniale ?' *Hérodote*, 135, La Découverte, 4e trim, 170–185.

Mooney, M. (2008). 'Structures de médiation et intégration des immigrants haïtiens à Paris', *REMI*, 24 (1), 89–114.

Moreau de Saint-Méry, M. L. E. (Philadelphia, 1797–1798). *Description topographique, physique, civile, politique et historique de la partie française de l'isle de Saint-Domingue*, Réédition: Tome I, II, III, 3ème édition, Paris: Publications de la Société française d'histoire d'outre-mer, 1565 p.

Morin, F. (1990). 'Des Haïtiens à NY: de la visibilité linguistique à la construction d'une identité caribéenne', in Simon-Barouth (ed.), *Les étrangers dans la ville. Le regard des sciences sociales*. pp. 340–355. Paris: L'Harmattan.

______. (1993). 'Entre visibilité et invisibilité: les aléas identitaires des Haïtiens de New York et Montréal', *REMI*, Poitiers, 9 (3), 147–176.

Morizot, F. (1988). *Grenade Epices et Poudre une épopée caraïbe*. Paris: L'Harmattan.

Morrison, D. (2012). *Reaching for the promised land: the role of culture, issues of leadership and social stratification within British Caribbean Christianity*, PhD Thesis, University of Birmingham, 359 p.

Morrisson, D. (2006). 'Growing Economic Dependence on the Diaspora', report, <http://www.jrs.net/news_detail?TN=NEWS-20120507060113>.

Mortley, N. K. (2009). 'Nurse Migration and The Impact on Health Systems in the Caribbean: the case of St. Lucia and Jamaica', in E. Thomas-Hope (ed.), *Freedom and Constraint in Caribbean Migration and Diaspora*, pp. 315–329. Kingston: Ian Randle Publishers,

Moya, J. C. (2005). 'Immigrants and Associations A Global and Historical Perspective', *JEMS*, 31 (5). 833–864.

Moyo, D. (2009). *L'aide fatale*. Paris: JC Lattès.

Munroe, T. (1972). *The Politics of Constitutional Decolonization: Jamaica 1944–1962*, Mona, Jamaica, ISER.

______. (1983). 'Quand l'oligarchie est au pouvoir …', *La Nouvelle revue internationale*, 26 (3), 96–104.

______. (1999). *Renewing Democracy into the Millenium: the Jamaican Experience in Perspective*, Mona: University of the West Indies.

Munroe, T., and Robotham, D. (1977). *Struggles of the Jamaican People*, Kingston, Workers' Liberation League.

Musée du Nouveau Monde (1985). *Mémoire d'un Port. La Rochelle et l'Atlantique XVIe-XIXe siècle*. La Rochelle: Imprimerie Rochelaise.

Myer, J. P. (1996). *Reformation in La Rochelle: Tradition and Change in Early Modern Europe, 1500–1568*. Geneva: Librairie Droz.

Nasso, Y. (1998). *La communauté haïtienne et son intégration en Guadeloupe*, Master's Thesis (Geography), Université des Antilles et de la Guyane.

Nedelcu, M. (2009). *Le migrant online. Nouveaux modèles migratoires à l'ère du numérique*. Paris: L'Harmattan (coll. Questions sociologiques).

Neill, G. (1986). *L'immigration haïtienne au Québec: insertion et trajectoire socio-professionnelle*, Master's Thesis, CRC, Université de Montréal.

Neill, G., and Kempeners, M. (1985). *Trajectoires socioprofessionnelles des immigrantes et immigrants haïtiens au Québec*. Université Montréal.

Nemours, A. (1962). 'The Alfred Nemours Collection of Haitian History: A Catalogue', Jane Toth (ed.), *Caribbean Studies*, 2 (3).

Nerestant, M. (1989). *Immigration haïtienne à Paris de 1825 à 1988*, Master's Thesis (Sociology), Institut Catholique de Paris.

Nettleford, R. (1979). *Caribbean Cultural Identity: the case of Jamaica*. Chicago: UCLA Press.

______. (2004). 'The Haitian Revolution and the Struggle against Slavery: Challenges to Knowledge, Ignorance and Silence ?' Colloque UNESCO.

Neveu, C., et al. (2007). 'Comparer en sciences sociales: aménagement et Anthropologie', in *Rapport final d'activités du réseau pluridisciplinaire de recherche*, pp. 114–121.

Neveu, C. (2009). 'Quelles questions pose "l'Autre" – devenu "le même" – à la majorité ? Interrogations sur la citoyenneté', in K. Akoka, P. Gonin and D. Hamelin (eds), *Migrants d'Ici et d'Ailleurs Du Transnational au local*, pp. 71–98. Poitiers: Atlantique.

Nicholls, D. (1979). *From Dessalines to Duvalier, Race, Color and National Independence in Haïti*. Cambridge: Cambridge University Press.

Nicolas, T. (2005). 'Le fragile équilibre d'une île-carrefour: Saint-Martin', in N. Bernadie and F. Taglioni, *Les dynamiques contemporaines des petits espaces insulaires*, pp. 163–180. Paris: Karthala, 2005.

Nora, P. (1989). 'Between memory and history: les lieux de mémoire'. *Représentations* 26: 7–24.

Noxolo, P. (2017). 'Introduction: Decolonising geographical knowledge in a colonised and re-colonising postcolonial world', *Area*, 49.3, 317–319, doi: 10.1111/area.12370.

N'zengou-Tayo, M.-J. (2011). 'La montagne ensorcelée: la promesse d'une Œuvre', in F.-A. Leconte (ed.), *Jacques Roumain et Haïti: la mission du poète dans la Cité*, pp. 39–51. Paris: L'Harmattan.

Offner, J.-M., and Pumain, D. (1996). *Réseaux et Territoires: significations croisées.* La Tour d'Aigue: éd. de l'Aube (coll. Territoire).

Ong, A. (1999). *Flexible Citizenship: The Cultural Logics of Transnationality.* Durham, NC: Duke University Press.

Oriol, M. (1996). 'Structure foncière, réforme agraire et agro-foncière: concepts et réalités', in Le développement rural en Haïti, *Conjonction* (200),160–162.

Orozco, M. (2000). Latino Hometown Associations as Agents of Development in Latin America, Inter-American dialogue Tomas Rivera, Washington, DC.

______. (2004). 'Understanding the remittance economy in Haiti', Paper Commissioned by The World Bank, Washington, DC.

______. (2005). 'Transnational Engagement, Remittances and Their Relationship to Development in Latin America and the Caribbean', *ISIM*. Georgetown University, Washington, DC.

Oscar, J. (2003). *Les immigrant(e)s haïtien (ne)s à Montréal et la perception de leur rôle dans le développement d'Haïti*, Master's Thesis, University of Ottawa.

Østergaard-Nielsen, E. (2001). 'Transnational political practices and the receiving state: Turks and Kurds in Germany-the Netherlands', *Global Networks* (1), 261–281.

Otok, S. (1988). *Research Approach to Social Geography*. Warszawa: University of Warsaw, 187 p.

Owen, D. (2001). 'Minority Ethnic Groups in Birmingham' in C. Lloyd, D. Jolyand and D. Owen (eds), *Blacks and Brittanity*, pp. 58–91. London: Ashgate.

Ozouf-Marignier, M.-V. (2009). 'Le territoire, la géographie et les sciences sociales: aperçus historiques et épistémologiques', in V. Martin, *Territoires, territorialité, territorialisation. Controverses et perspectives*, pp. 20–31. Rennes: PUR.

Palmer, J. L. (2008). *Atlantic crossings: Race, gender, and the construction of families in 18th century*, PhD Thesis, The University of Michigan.

Palmer, R. W. (2009). 'The Remittance Profile of Jamaican Immigrants', in Thomas-Hope (ed.), *Freedom and Constraint in Caribbean Migration and Diaspora*, pp. 258–279. Kingston: Ian Randle Publishers.

Paret, L. V. (1962). 'Quelques aspects de la main-d'œuvre haïtienne', *Journal of Inter-american Studies*, 4 (1).

Pastor, R. (1985). 'Migration and Development in the Caribbean: Relating Policies and People', *International Migration Review* 19 (1), 144–151.

Patterson, P. J. (2010). *Migration and Development in the Commonwealth: A Caribbean Perspective*, The Walter Rodney Memorial Lecture. The University of Warwick.

Paul, B. (2008). *Migration et pauvreté en Haïti: impacts économiques et sociaux des envois de fonds sur l'inégalité et la pauvreté*, Univ. Bordeaux.

Payne, A., and Sutton, P. (2001). *Charting Caribbean Development*. Gainsville: University Press of Florida.

Peach, C. (1991). *The Caribbean in Europe: contrasting patterns of migration & settlement in Britain, France and the Netherlands* (15), Warwick, CRER.

______. (2007). 'Sleepwalking into Ghettoisation: the British Debate over Segregation', in K. Schönwälder (ed.), *Residential Segregation & the Integration of Immigrants: Britain, the Netherlands & Sweden*, SP IV 602.

Péan, L. (2000). *Économie politique de la corruption (de Saint-Domingue à Haïti 1791–1870)*, Port-au-Prince: Éditions Mémoire, 523 p.

______. (2009). *Aux origines de l'État marron en Haïti (1804–1860)*. Port-au-Prince: Éditions de l'Université d'État d'Haïti, 391 p.

Pecqueur, B. (2000). *Le développement local pour une économie des territoires*. Paris: Syros (coll. Alternatives économiques).

Perchellet, S. (2010). *Haïti Entre colonisation, dette et domination, 2 siècles de luttes pour la liberté*. Liège: Editions CADTM.

Peretz, H. (2002). *Les méthodes en sociologie, L'observation*. Paris: La Découverte.

Perouse de Montclos, M-A. (2015). Pour un développement 'humanitaire' ? Les ONG à l'épreuve de la critique. Paris: IRD Editions (Coll. Objectifs Suds).

Petit, V., Desbois, C., and Charbit, Y. (1999). 'De la difficulté d'être passeur de francophonie pour les migrants portugais dans leur pays d'origine', *Rapport CERPAA*.

Phelps, A. (2006, 1st edn, 1966). *Mon Pays que voici*. Montréal: Mémoire d'encrier.

Phillip-Dowe, N. (2013). 'Women in Grenada Revolution, 1979–1983', *Social Economic Studies* 62, 45–82.

Phillips, A. (2008). Haiti, France and the Independent Debt of 1825, paper given at the University of San Francisco Law School, co-sponsored by the Institute for Justice and Democracy in Haiti

Philoctète, R. (1998). *Ces îles qui marchent* (coll. Spirale). Port-au-Prince: Edition Mémoire.

Pialou, A. (2007). 'La Guyane ou le "tout reconduite"', *Plein Droit*, GISTI, 74.

Piantoni, F. (2002). 'Les recompositions territoriales dans le Maroni: relation mobilité environnement', *REMI*, 18 (2), 11–49.

______. (2009). *L'enjeu migratoire en Guyane française*. Paris: Ibis Rouge Ed.

Pierre, B. (2006). *L'insertion des Haïtiens et des Saint-Luciens en Martinique*, Master's Thesis (Geography), Université des Antilles et de la Guyane.

Pierre, S. (2007). *Ces Québécois venus d'Haïti. Contribution de la communauté haïtienne à l'édification du Québec moderne*. Montreal: Presses internationales Polytechnique.

______. (ed.) (2010). *Construction d'une Haïti nouvelle Vision et contribution GRAHN*. Québec: Presses Internationales Polytechnique.

Pierre-Louis, F., Jr (2001). *Transnational Organizations and Citizen Participation: A Study of Haitian Immigrants in New York City*, CUNY.

Pierre-Paul, E. (2001). *Saint-Marc, je m'en souviens*. Québec: CIDIHCA (coll. Libre Pensée).

Plaza, D. (2009). 'Transnational Return Migration to the English Speaking Caribbean', in E. Thomas-Hope (ed.), *Freedom and Constraint in Caribbean Migration and Diaspora*. Kingston: Ian Randle Publishers.

Plaza, D., and H. Frances (2006). *Returning to the Source The Final Stage of the Caribbean Migration Circuit*. Jamaica: University of the West Indies Press.

Poinsot, M. (2001). 'Le mouvement associatif, un instrument au service des politiques publiques d'intégration ?', in 'Vie associative. Action citoyenne', *Hommes & Migration*, 1229 (2), 64–69, 71–73.

Portes, A. (1998). 'Social Capital: Its origins and applications in modern Sociology', *Annual Review of Sociology*, 24, 1–24.

Portes, A., and de Wind, J. (2008). *Rethinking migration: new theoretical and empirical perspectives*. Oxford: Berghahn Books.

Portes, A., Haller, W., and Guarnizo, L. E. (2002). 'Les entrepreneurs transnationaux: une forme alternative d'adaptation économique des immigrés', in M. Péraldi (ed.), *La fin des norias ? Réseaux migrants dans les économies marchandes en Méditerranée*, pp. 51–89. Maisonneuve.

Portes, A., and Landolt, P. (1996). 'The Downside of the Social Capital', *The American Prospect*, 7 (26).

Portes, A., and Rumbaut, R. G. (1996). *Immigrant America: A Portrait*. University of California Press.Pouligny, B. (2004). *Ils nous avaient promis la paix*. Paris: Les presses de Sciences Po.

Portes, A., et al. (2007). 'Immigrant Transnational Organizations and Development: A Comparative Study', *IMR*, 41 (1), 242–281.

Préval, C. J. (1999). *Marmelade, mon amour, mon tourment*! Pages retrouvées. Port-au-Prince.

Price, R. (1983a). *Maroon Societies. Rebel Slave Communities in the Americas*. Baltimore, MD: Johns Hopkins University Press.

Price-Mars, J. (2009). *Ainsi parla l'Oncle suivi de Revisiter l'Oncle* (réédition de l'essai, 1927). Montréal: Mémoire d'Encrier.

Privert, J. (2006). *Décentralisation et Collectivités Territoriales (contraintes, enjeux et défis)*. Port-au-Prince: Editons le béréen.

Proulx, M. U. (1995). *Réseaux d'information et dynamique local.* Chicoutimi, GRIR.

Prujiner, A. (1994). 'Nationalité, migrations et relations internationales', in B. Badie and C. Wihtol de Wenden (eds), *Le défi migratoire Questions de relations internationales*, pp. 105–126. Presses F. Nationales.

Puri, S. (2003). 'Introduction: Theorizing Diaspora Cultures: The Quiet Migrations', in *Marginal Migrations: The Circulation of Cultures within the Caribbean,* pp. 1–16. London: Macmillan.

Putnam, R. (1993). *Making Democracy Work: Civic Traditions in Italy*, Princeton, NJ: Princeton University Press.

Przeworski, A., et al. (1996). 'What Makes Democracies Endure?' *Journal of Democracy*, 7 (I), 39–55.

Quiminal, C. (1991). *Gens d'ici, gens d'ailleurs*. Paris: Christian Bourgeois.

Raffestin, C. (1980). *Pour une géographie du pouvoir*. Paris: Librairies technique (coll. Géographie économique et sociale).

Ratia, C. (2001). *La communauté haïtienne de Montréal: immigration, insertion dans l'espace urbain et dynamique actuelle*, Master's Thesis, Bordeaux III.

______. (2002). *Haïti et sa diaspora depuis le début des années 1980*, Master's Thesis DEA, Université de Poitiers.

Raymond, G., and Modood, T. (eds) (2007). *The Construction of Minority Identities in France and Britain*. Seattle, WA: Magus Books Seattle.

Redon, M. (2009). 'Migrations et frontière: le cas de Saint-Martin', *Études caribéennes*, 8 <http://etudescaribeennes.revues.org/>.

Reno, F. (2003). 'Pour une approche sociohistorique du politique dans les Caraïbes', in J. A. Lafontant, *Participation citoyenne, Gouvernance décentralisée et Démocratie participative,* Colloque international, PNUD, Port-au-Prince.

______. (2008). 'L'immigrant haïtien, entre persécutions et xénophobies', *Hommes et Migrations*, 1274, 132–142.

______. (2012). *Mobilisations sociales aux Antilles*, J.-C. William, F. Reno F. and Alvarez (eds). Paris: Karthala.

Retaillé, D. (1997). *Le monde du géographe*. Presses de Sciences Po.

Rex, J., and Tomlinson, S. (1979). *Colonial Immigrants in a British City: A Class Analysis*. London: Routledge.

Ribbe, C. (2005). *Le crime de Napoléon*. Paris: Editions Privé.

Richardson, D. Schwarz, S. and Tibbles, A. (2007). *Liverpool and Transatlantic Slavery*. Liverpool: Liverpool University Press.

Rist, G. (1996). *Le développement, histoire d'une croyance occidentale*, Paris: Presses de Sciences Po (coll. Références inédites).

Robinson, M. (2013). *Mary Robinson. Everybody Matters*. London: Holder.

Rodrik, D. (2003). *In Search of Prosperity*. Princeton, NJ: Princeton University Press.

Romain, G. (2005). *Getting into Local Power: the Politics of Ethnic Minorities in British and French Cities*. Oxford: Blackwell Publishing, 266 p.

Rostow, W. (1962). *Les étape de la croissance économique*. Paris: Editions Seuil.

Roumain, J. (1942). *Contribution à l'étude de l'ethnobotanique précolombienne des Grandes Antilles*, Port-au-Prince, Imprimerie de l'état.

Roy, O. (2008). *La Sainte Ignorance*. Paris: Editions du Seuil.

Sabatier, P. A. (2007). *Theories of the Policy Process*, 2nd edn. Boulder, CO: Westview Press.

Saïd, E. (1978). *L'orientalisme L'Orient créé par l'Occident*. Paris: éditions du Seuil.

Saint-Hubert, F. (2003). *Et le neveu répondit. Analyse statistique de la migration haïtienne aux États-Unis (1953–2000)*, Victoria (Canada): Trafford Publishing.

Saint-Lot, E. (1956). 'L'impasse de la Déclaration des droits de l'homme', Revue optique, rubrique Société, *Le Nouvelliste* (juillet), 7p.

Saint-Louis, D. (1988). *L'immigration haïtienne à Paris*, Institut Catholique de Paris, Master's Thesis.

Saint Pierre, M. (2015). *Eric Williams and The Anticolonial Tradition. The Making of a Diasporan Intellectual*. University of Virginia Press.

Saint-Vil, J. (2011). 'Les Haïtiens déportés des États-Unis', L.-A. Joint and J. Mérion (eds), *L'immigration haïtienne dans la Caraïbe*, pp. 263–279. Gourbeyre (Guadeloupe): Éditions Nestor.

Saintilus, P. N. (2007). 'Ouanaminthe et la dynamique transfrontalière Haïtiano-dominicaine', *Dynamiques migratoires de la Caraïbe*, coll. Terres d'Amérique, pp. 323–343. Karthala/Géode Caraïbe.

Salas-Molins, L. (1987). *Le Code Noir ou le calvaire de Canaan*. PUF.

Salome, G. (2011). 'Camps, *Transitional Site* et Relogement à Port-au-Prince après le séisme du 12 janvier', 10 p. <http://www.sciencespo.fr/ceri/sites/sciencespo.fr/art_gs.pdf>.

Sassen, S. (2002). *De-Nationalization*. Princeton, NJ: Princeton University Press.

Schnapper, D. (2001). 'De l'Etat-nation au monde transnational. Du sens et de l'utilité du concept de diaspora', *REMI*, 17 (2), 9–36.

Schoburgh, E. D. (2003). 'Modernising local government by fragmentation: Lessons from the Portmore municipal experiment, Jamaica', *Commonwealth Journal of Local governance*, Special issue, 24 p.

Seitenfus, R. (2015). *L'échec de l'aide internationale à Haïti. Dilemmes et Egarements*. Preface Peck. Postface C. Hector. Port-au-Prince: Editions Université d'Etat d'Haïti.

Sertima, I. V. (1976). *They came before Columbus*. New York: Random House.

______. (2006). 'The Place of Africa in History and Scholarship: A conversation with Ivan Sertima', in Edward Ramsamy (ed.), *Conceptualizing Race, Empire, Modernity and Representation*, Chapter II, pp. 8–14.

Shain, Y., and Barth, A. (2003). 'Diasporas and International Relations Theory', *International Organization*, 57 (3), 449–479.

Sheller, M. (2000). *Democracy after Slavery Black Politics and Peasant Radicalism in Haïti and Jamaica*. Gainesville: University Press of Florida.

______. (2012). *Citizenship from Below*, Durham, NC: Duke University Press, 346 p.

Shellman, S., and Stewart, B. (2007). 'Political Persecution or Economic Deprivation? A Time-Series Analysis of Haitian Migration to the United States', *Conflict Management & Peace Science*, 24 (2), 121–137.

Shepherd, V. 'Reparation & the Right to Development', paper delivered at the 16th session of the *WGEPAD*, Geneva, 30 March-2 April 2015.

Simon, G. (1995). *Géodynamique des migrations internationales dans le Monde*. Paris: PUF.

______. (2006). 'Migrations, la spatialisation du regard', *REMI*, 22 (2), 9–21.

______. (2009). 'Les migrations internationales dans le mouvement du monde', in *Migrants d'Ici et d'Ailleurs Du transnational au Local*, Karen Akoka, P. Gonin et D. Hamelin (eds), Atlantique, pp. 71–98.

Sives, A., Morgan, J., and Appleton, S. (2006). 'Teacher Migration from Jamaica: Assessing the short-term impact', *Caribbean Journal of Education*, 27 (1), 85–166.

Skeldon, R. (2006). 'Interlinkages between internal and international migration and Development in the Asian region', *Population, Space and Place*, 12, 15–30.

______. (2010). 'Migration & development over 20 years of research: progress & prospects', in C. Audebert and K. M. Doraï (eds), *Migration in a Globalized World*, pp. 145–158. Amsterdam: Amsterdam University Press.

Skelton, T. (2004). *Introduction to the Pan-Caribbean*. Oxford: Oxford University Press.

Slater, D. (1976). 'Critique de la géographie du sous-développement', vol. LX, pp. 59–96. *Cahiers internationaux de sociologie*.

Smith, M. J. (2004). 'Emperor, Exiles, and Intrigues: The Case of Nineteenth-Century Haitian Heads of State in Jamaica', in *The Haiti-Jamaica Connection*, pp. 35–43. Kingston: Latin American-Caribbean Centre, UWI.

______. (2011). 'From the Port of Princes to the City of Kings: Jamaica and the Roots of the Haitian Diaspora', in Regine O. Jackson (ed.), *Geographies of the Haitian Diaspora*, New York: Routledge, pp. 17–33.

______. (2017). 'Loving and Leaving the New Jamaica: Reckoning with the 1960s', keynote speech, 'Memory, Migration, and Decolonisation in the Caribbean and Beyond, 1804 to the Present', ILAS, Senate House, London, UK.

Smouts, M-C. (2007). *La situation post-coloniale: les postcolonial studies dans le débat français*. Paris: FNSP.

Souffrant, C. (1970). 'Travailleurs haïtiens en République dominicaine', *Migrations dans le monde*, n° 3 (juillet), pp. 1–4.

______. (1974). 'Les Haïtiens aux États-Unis', *Population*, Paris, 29 (1), 133–145.

Soukar, M. (2016). *Il y a deux cents ans Simon Bolivar aux Cayes 1815/1816–2015/2016*. Port-au-Prince: C3 Editions.

Soysal, Y. (1994). *Limits of Citizenship: Migrants and Postnational Membership in Europe*. Chicago: University of Chicago Press.

Spickard, J. (1999). The Origins of the Universal Declaration of Human Rights, University of Redlands, <http://bulldog2.redlands.edu/fac/spickard/onlinepubs/originudhr.pdf>.

Supplice, D. (2009). *De la naturalisation en Haïti*. Port-au-Prince: Deschamps.

______. (2017). *Catalogue de l'Exposition de documents historiques haïtiens. Collection Daniel Supplice*. Ambassade de la république d'Haïti près le Royaume de Belgique. Les Musées royaux d'art et d'histoire de Bruxelles (18 mai).

Tapia, S. de (1998). *La circulation migratoire dans les diasporas ouvrières de la Méditerranée. Turquie Maroc*. Rapport final d'une recherche financée par le Ministère de l'éducation nationale et de la culture, Poitiers, MIGRINTER.

Tardieu, J. (2005). *L'avenir en face Haïti à l'épreuve de la mondialisation et du défaitisme de ses élites*. Montréal: les Editions du CIDIHCA.

Tarrius, A. (1996). 'Territoires circulatoires des migrants et espaces européens' in M. Hirshborn et J. M. Berthelot (eds), *Mobilités et ancrages. Vers un nouveau mode de spatialisation ?* Paris: L'Harmattan.

______. (2001). 'Au-delà des Etats-nations: des sociétés de migrants', *REMI*, 17 (2), 37–61.

Théodat, J-M. (2003). *Haïti-République dominicaine. Une île pour deux: 1804–1916*. Paris: Karthala.

______. (2007). 'Haïti et la République Dominicaine: la négritude en partage', in *Africultures Diaspora: identité plurielle*, L'Harmattan, 72, 36–45.

______. (2008). 'L'autre pays en dehors: le pays lotbodlo', *revue Conjonction*, N° spécial Gérard Barthélémy, pp. 39–52, Institut français d'Haïti.

______. (2012). 'Au-delà de la Frontière, les Relations Haïti – République dominicaine à l'épreuve du tremblement de terre du 12 janvier 2010', *Rencontres*, 24–25 (février), pp. 23–27, CRESFED.

Therwath, I. (2007). *L'Etat face à la Diaspora: stratégies et trajectoires indiennes*, PhD Thesis, IEP, Paris.

Thiollet, H. (2007). *Migrations et intégrations dans le sud de la mer Rouge: migrants et réfugiés érythréens au Soudan, au Yémen et en Arabie Saoudite 1991–2007*, PhD Thesis. Institut d'études politiques de Paris.

______. (2011). 'Migration as Diplomacy: Labor Migrants, Refugees, and Arab Regional Politics in the Oil-Rich Countries', *International Labor and Working-Class History*, 79, Spring,103–121.

Thomas, D. A. (2004). *Modern Blackness, Globalization and the Politics of Culture in Jamaica*, Durham, NC: Duke University Press.

Thomas-Hope, E. (ed.) (1983). 'Caribbean identity a matter of perception', in *Perspectives on Caribbean Regional Identity*, pp. 1–4. Liverpool: CLAS

______. (1992). *Explanation in Caribbean Migration Perception and the Image: Jamaica, Barbados, St Vincent*, London: MacMillan Press LTD.

______. (1996). *The Environmental Dilemma in Caribbean Context*, Kingston: Grace Kennedy Foundation Lecture.

______. (1999). 'Return Migration to Jamaica and its Development Potential', *IMR*, vol. 37 (1), pp. 183–202.

______. (2002). 'Transnational livelihoods and identities in return migration to the Caribbean: The case of skilled returnees to Jamaica', in N. Sorensen and K. F. Olwig, (eds), *Work and Migration: Life and Livelihoods in a Globalizing World*, pp. 187–201. London: Routledge,

______. (2009). *Freedom and Constraint in Caribbean Migration and Diaspora*, Kingston: Ian Randle Publishers.

Thomas, P., and Bauer, E. (2000). 'Jamaican Transnational Families: Points of pain and Sources of Resilience', *Wadabagei: A Journal of the Caribbean and its Diaspora*, 3 (2), 1–36.

Thomas-Hope, E., and Nutter, R. D. (1989). 'Occupation and Status in the Ideology of Caribbean Return', in Reginald Appleyard (ed.), *The Impact of International Migration on Developping Countries*, Paris: OECD publication.

Thurmes, M. (2006). *Les Métropolitains en Guyane: Une intégration sociale entre individu et groupe culturel*, PhD Thesis, Université Montpellier III.

Todd, E. (2015). *Qui est Charlie, Sociologie d'une crise religieuse*. Paris: Seuil.

Tolia-Kelly, D. (2006) Affect – an ethnocentric encounter? Exploring the 'universalist' imperative of emotional/affectual geographies. *Area* 38(2): 213–217.

______. (2016). Feeling and Being at the (Postcolonial) Museum: Presencing the Affective Politics of 'Race' and Culture, *Sociology* 50: 1–17.

Toulis, N. R. (1997). *Believing Identity: Pentecostalism and the Mediation of Jamaican Ethnicity and Gender in England*. Oxford: Berg.

Tourme-Jouannet, E. (2013). 'Reparations for Historical Wrongs: The Lessons of Durban', in Chapter 8, *What is a Fair International Society?* Hart Publishing, Oxford, pp. 187–201.

Toussaint, H. (ed.) (2009). *Haïti et la France 2003–2004 un débat escamoté*, Collectif des universitaires-citoyens (CUCI), Presses Média-texte.

Traoré, Aminata Dramane (2014). *La gloire des imposteurs. Lettres sur le Mali et l'Afrique*. Edition Philippe Rey.

Tremblay, D. G., and Fontan, J.-M. (1994). *Développement économique local: la théorie,la pratique, les expériences*, Presses Université du Québec.

Trouillot, L. (2001). 'Quelle mémoire pour nous ?' in *Haïti (re) penser la citoyenneté*, Collection Pour l'Etat de droit en Haïti. Port-au-Prince: les Editions HSI, vol. VIII.

Trouillot, M.-R. (1977). *Ti dife boule sou Istwa Ayiti*, Edysion KIK, Inivèsite Karayib.

______. (1983). 'The Production of Spatial Configurations: A Caribbean case', *New West Indian Guide*, 57 (3/4), 215–229.

______. (1986). *Les racines historiques de l'Etat duvalérien*. Port-au-Prince: Deschamps.

______. (1995). *Silencing the Past: Power and the Production of History*. Beacon Press.

Turnier, A. (1989). *Quand la nation demande des comptes*. Port-au-Prince: Le Natal.

Urfié, J. Y. (1986). 'Les Haïtiens en Guyane, une immigration en cours de stabilisation ?', In *Les dossiers de l'Outre-mer*, 85 (4), 28.

Vachon, B., and Coallier, F. (1993). *Le développement local: théorie et pratique. Réintroduire l'humain dans la logique du développement local*, Boucherville: Gaëtan Morin.

Vairac, M. (2008). ' La question de l'immigration dans le discours politique, le cas des Haïtiens en Rép. Dominicaine', *Hommes et migrations*, 1274, 48–59.

Vallès, M., et al. (2013). *Lettre à Toi Notre Papa-Bon-Dieu*, Combite Quisqueyenne.

Valmé, G. R. (2012). *Atabey, Yucayequey, Caney, 6 000 ans d'aménagement territorial préhispanique sur l'île d'Ayiti Haïti/ République Dominicaine*. Miami: Educa Vision.

Vargas-Sylva, C. (2012). *The handbook of Research Methods in Migration*. Northhampton, MA: Oxford University, 608 p.

Vastey, Pompée-Valentin, Baron de (1814). *Le système colonial dévoilé*. Re-edition by the Société haïtienne d'Histoire, de géographie et de géologie (SHGG) Imprimerie Lakay (2013).

______. (1823). *An Essay on the Causes of the Revolution and Civil Wars of Hayti, Journals Concerning Hayti*. Exeter, England: Western Luminary Office.

Vaughan, G., et al. (2010). *Le monde britannique (1815–1931)*. Paris: Ed. Belin.

Verges, F. (2005). 'Postface: Pour une lecture postcoloniale de Césaire', in Aimé Césaire, *Nègre je suis, nègre je resterai*, Albin Michel, 71–136.

Verteuil de, M. (1995). *Envol pour les abricots: récit d'une aventure de développement dans une région rurale d'Haïti*, Montréal: T.P.F.

Vigour, C. (2005). *La comparaison dans les sciences sociales Pratiques et Méthodes*. Paris: La Découverte.

Voyneau, S. (2005). 'République Dominicaine: le traitement infligé aux Haïtiens et aux Dominicains d'origine haïtienne, une discrimination institutionnalisée ?'. *La chronique des Amériques*, Montréal, 33.

Wa Thiong'o, N. (2011). *Decolonising the mind: the politics of language in African literature*. Oxford: James Currey.

Wah, T. (2003). *Haiti's Development through Expatriate Reconnection*. Florida: EducaVision.

Wainwright, J. (2004). *Haïti Agenda Information*. Port-au-Prince: Deschamps.

Waldinger, R. (1993). 'Le débat sur l'enclave ethnique: revue critique', *REMI*, 9 (2), 15–29.

______. (2006). 'Transnationalisme des immigrants et présence du passé', *REMI*, 22 (2), 23–41.

Waldinger, R., and Fitzgerald, D. (2004). 'Transnationalism in Question', *American Journal of Sociology (AJS)*, 109 (5), 1177–1195.

Weil, P. (1988). *L'analyse d'une politique publique. La politique française d'immigration 1974–1988*, PhD Thesis, IEP Paris.

Weil, P., and Hansen, R. (eds) (1999). *Nationalité et citoyenneté en Europe*. Paris: La Découverte.

Weiss, C. H. (1998). *Evaluation*, 2nd edn. Upper Saddle, NJ: Prentice Hall.

Weiss Fagen, P., et al. (2010). *Diasporas in Conflict Zones: An Assessment of Neglected Agents of Change*, Report, Institute for the Study of International Migration, Georgetown University.

Werner, M. (2016). *Global Displacements. The Making of Uneven Development in the Caribbean*. West Sussex: Wiley Blackwell.

Wieviorka, M. (1998). 'Racisme, antiracisme et mutation sociale L'expérience Française', in Andréa Rea (ed.), *Immigration et racisme en Europe*, Editions Complexe, 23–44.

______. (2007). 'The Construction of What?' in G. Raymond and T. Modood, *The Construction of Minority Identities in France & Britain*, pp. 33–47. Seattle, WA: Magus Books Seattle.

Wihtol de Wenden-Didier, C. (1986). *Les immigrés et la politique, thèse de doctorat d'Etat*, Institut d'études politiques de Paris.

______. (1987). *Citoyenneté Nationalité et Immigration*. Paris: Arcantères (Coll. Migrations plurielles).

______. (2001). *La Beurgeoisie Les trois âges de la vie associative issue de l'immigration*. Paris: CNRS Editions.

______. (2002). 'La citoyenneté européenne', in 'Nationalité et citoyenneté, nouvelle donne d'un espace européen', Travaux du CEP (mai).

______. (2009). *Atlas mondial des migrations: Réguler ou réprimer ... gouverner*. Cartographie de Madeleine Benoit-Guyod. Editions Autrement.

______. (2010). *La question migratoire au XXIe siècle Migrants, réfugiés et relations Internationales*. Paris: les Presses de Sciences Po.

Wihtol de Wenden, C., and Grandmaison Le Cour, O. (1993). *Les étrangers dans la cité expériences européennes.* Paris: La Découverte.

Wilberforce, R. and Samuel (eds) (1840). *The Correspondence of William Wilberforce*, Vol. 1. Albemarle Street, London: John Murray.

William, E. (1975). *De Christophe Colomb à Fidel Castro: l'histoire des Caraïbes 1492–1969.* Paris: Présence Africaine.

William, J.-C., Reno, F., and Alvarez, F. (2012). *Mobilisations sociales aux Antilles.* Paris: Karthala.

Williams, S., and Lew, A. (2015). *Tourism Geography.* Oxon, New York: Routledge.

Winston, J., and Harris, C. (1993). *Inside Babylon: Caribbean Diaspora in Britain.* London and New York: Verso Books.

Wolton, D. (2006). *Demain la francophonie.* Paris: Flammarion.

Wresinski, J. (2007). *Refuser la misère Une pensée politique née de l'action.* Paris: Les éditions du Cerf.

Wynter, C. (2001). *Jamaican Labor Migration to Cuba, 1885–1930, in the Caribbean Context*, PhD Dissertation, University of Illinois Chicago.

Ying, N., et al. (2010–2011). Jamaica Diaspora Institute, *Inventory Research to Diaspora and Migration*, UWI, Mona Campus.

Young, R. (2003). *Post-Colonialism: A Very Short Introduction.* Oxford: Oxford University Press.

Zacaïr, P. (2005). 'Haiti on his mind: Antonio Maceo and caribbeanness', *Caribbean Studies* (Universidad de Puerto Rico), 33 (1), 47–78.

______. (2010). *Haiti and the Haitian diaspora in the wider Caribbean.* Gainesville: University Press of Florida.

Zephyrin, R. (2005). *Le champ migratoire haïtiano-guyanais: étude des causes et effets politiques, socio-économiques et spatiaux.* PhD Thesis, Université d'Aix-en-Provence.

______. (2012). 'Transferts d'argent des migrants haïtiano-guyanais dans la multipolarité de l'espace américain', *Migrations Société*, 24 (143), 33–50.

Zinn, H. (1980). *Une Histoire populaire des Etats-Unis de 1492 à nos jours* (traduit de l'anglais par Frédéric Cotton), Marseille: Agone.

Zolberg, A. (2006). Colloque Compte-rendu Sciences Po-CERI (6 octobre), Presentation 'A Nation by Design, Immigration policy in the fashioning of America', The Russell Sage Foundation.

Index

CULTURAL MEMORIES

SERIES EDITOR

Katia Pizzi
Director, Centre for the Study of Cultural Memory
Institute of Modern Languages Research, University of London

Cultural Memories is the publishing project of the Centre for the Study of Cultural Memory at the Institute of Modern Languages Research, University of London. The Centre is international in scope and promotes innovative research with a focus on interdisciplinary approaches to memory.

This series supports the Centre by furthering original research in the global field of cultural memory studies. In particular, it seeks to challenge a monumentalizing model of memory in favour of a more fluid and heterogeneous one, where history, culture and memory are seen as complementary and intersecting. The series embraces new methodological approaches, encompassing a wide range of technologies of memory in cognate fields, including comparative studies, cultural studies, history, literature, media and communication, and cognitive science. The aim of *Cultural Memories* is to encourage and enhance research in the broad field of memory studies while, at the same time, pointing in new directions, providing a unique platform for creative and and forward-looking scholarship in the discipline.

Vol. 1 Margherita Sprio
Migrant Memories: Cultural History, Cinema and the Italian Post-War Diaspora in Britain
2013. ISBN 978-3-0343-0947-9

Vol. 2 Shanti Sumartojo and Ben Wellings (eds)
Nation, Memory and Great War Commemoration: Mobilizing the Past in Europe, Australia and New Zealand
2014. ISBN 978-3-0343-0937-0

Vol. 3 Cara Levey
Fragile Memory, Shifting Impunity: Commemoration and Contestation in Post-Dictatorship Argentina and Uruguay
2016. ISBN 978-3-0343-0987-5

Vol. 4 Katia Pizzi and Marjatta Hietala (eds)
Cold War Cities: History, Culture and Memory
2016. ISBN 978-3-0343-1766-5

Vol. 5 Renata Schellenberg
Commemorating Conflict: Models of Remembrance in Postwar Croatia
2016. ISBN 978-3-0343-1901-0

Vol. 6 Stephen Wilson and Deborah Jaffé (eds)
Memories of the Future: On Countervision
2017. ISBN 978-3-0343-1935-5

Vol. 7 Patrizia Violi
Landscapes of Memory: Trauma, Space, History
2017. ISBN 978-3-0343-2202-7

Vol. 8 Samuel Merrill
Networked Remembrance: Excavating Buried Memories in the Railways beneath London and Berlin.
2017. ISBN 978-3-0343-1919-5

Vol. 9 Dirk Göttsche (ed.)
Memory and Postcolonial Studies: Synergies and New Directions.
Forthcoming. ISBN 978-1-78874-478-2

Vol. 10 Clara Rachel Eybalin Casséus
Geopolitics of Memory and Transnational Citizenship: Thinking Local Development in a Global South
2019. ISBN 978-1-78707-978-6